Laser Communication with Constellation Satellites, UAVs, HAPs and Balloons

Arun K. Majumdar

Laser Communication with Constellation Satellites, UAVs, HAPs and Balloons

Fundamentals and Systems Analysis
for Global Connectivity

 Springer

Arun K. Majumdar
San Diego, CA, USA

ISBN 978-3-031-03974-4   ISBN 978-3-031-03972-0   (eBook)
https://doi.org/10.1007/978-3-031-03972-0

© Springer Nature Switzerland AG 2022
This work is subject to copyright. All rights are reserved by the Publisher, whether the whole or part of the material is concerned, specifically the rights of translation, reprinting, reuse of illustrations, recitation, broadcasting, reproduction on microfilms or in any other physical way, and transmission or information storage and retrieval, electronic adaptation, computer software, or by similar or dissimilar methodology now known or hereafter developed.
The use of general descriptive names, registered names, trademarks, service marks, etc. in this publication does not imply, even in the absence of a specific statement, that such names are exempt from the relevant protective laws and regulations and therefore free for general use.
The publisher, the authors and the editors are safe to assume that the advice and information in this book are believed to be true and accurate at the date of publication. Neither the publisher nor the authors or the editors give a warranty, expressed or implied, with respect to the material contained herein or for any errors or omissions that may have been made. The publisher remains neutral with regard to jurisdictional claims in published maps and institutional affiliations.

This Springer imprint is published by the registered company Springer Nature Switzerland AG
The registered company address is: Gewerbestrasse 11, 6330 Cham, Switzerland

This book is dedicated to the students, researchers, and entrepreneurs who might be inspired by the opportunity and challenge of the unlimited possibilities created by optical wireless communications that provides global access to all.

Foreword

Many space laser communications missions have been demonstrated and verified in orbit so far. The first in-orbit laser communication experiment with just 1 Mbit/s was conducted between a geostationary Earth orbit (GEO) satellite and an optical ground station in 1994. Since then, the data rate has been increased up to several Gbit/s, and more than 30 satellites with laser communication terminals have been launched in the last three decades from all over the world. In the early 1990s, space laser communication missions were conducted only by space agencies and governmental institutions; however, the private sector has been able to play a more relevant role in space in the recent years, which means that space laser communication technologies have become more mature, entering into the implementation phase within the information society, closing the gap between the research and development phase and commercialization. Recently, drones, UAVs, HAPs, and low Earth orbit (LEO) constellations have become hot topics, and the spectral resource of radio frequency (RF) will be exhausted under the environments of many flying objects in the future. The frequency allocation issue will be one of the biggest problems in Beyond 5G era. Toward Beyond 5G, the ultra-broadband communication will be an essential driving force for several Beyond 5G applications. The non-terrestrial networks (NTN) will play a major role in Beyond 5G era. In the 3D network applications, new services such as automatic logistics with drones, autonomous ships, global disaster monitoring with IoT satellites, and lunar broadband communications are expected among many others, yet to come. In 2030s, the XR (X-Reality) technology with ultra-broadband wireless communications will enable to realize future telework, remote medicine under Covid-19-like situations, ultrarealistic online gaming, virtual shopping with avatars, remote education, autonomous flying vehicles, ultrasmart society, etc. Free-space optical communications are inevitable for such ultra-broadband wireless networks, and they can also guarantee information security with quantum cryptography. There are some drawbacks in free-space laser communications in the atmosphere, which are the signal degradation due to atmospheric turbulence and the influence of weather conditions. Professor Arun K. Majumdar is an outstanding researcher and lecturer on atmospheric turbulence effects on free-space laser communications, propagation, and imaging. He timely writes this book entitled "Laser Communication with Constellation Satellites, UAVs, HAPs and Balloons for Global Connectivity: Fundamentals and Systems Analysis." This is an essential book for the future research and development of space laser communications and their implementation in the society. He deals with the subject exhaustively across theory, fundamentals, systems, technologies, and applications. This book covers the basics of satellite communications, laser beam propagation, satellite laser communications, optical link design, and integrated all-optical networks aiming for constellation of satellites. I truly recommend this book to the students for their education, and I hope that many young readers will be able to catch the actual significance and impact of these technologies to become the next generation of players to create the future space laser communication applications that the society will be able to enjoy.

Wireless Networks Research Center   Morio Toyoshima
National Institute of Information and Communications
Technology (NICT)
Koganei, Tokyo, Japan

Preface

Optical wireless communication (OWC) is an emerging area where continuous interests and active research are growing worldwide. The reasons for this are the tremendous future demand for high-bandwidth and secure communication for a number of communication applications driven by the ever-increasing number of wireless broadband Internet, mobile phones, smart devices, social networks, and video gaming. There is thus a growing increase in the data traffic due to the widespread use of smart devices anytime, anywhere, and consequently the volume of mobile and wireless users and the data traffic are predicted to increase thousandfold over the next decade. RF-based technologies and already existing wireless communication systems cannot solve the future high-data-rate worldwide communication and connectivity problems. Recent developments of laser satellites provide a future potential solution to establish high-data-rate global connectivity and communication using space optical networks, which include low Earth orbit (LEO) and geosynchronous orbit (GEO) satellites, aerial high-altitude platforms (HAPs) and UAVs, small satellites such as CubeSats, and terrestrial and underwater optical terminals. Free-space optics (FSO) referring to OWC play an important role in the creation of next-generation satellite networks. Furthermore, the proposed hundreds and thousands of constellation satellites in different orbits for establishing global communication and connectivity to provide communication services to undeserved and remote areas are subjects of current research and development.

The motivation of creating such a book started when the author was invited to be a panelist at the Optical Society of America (OSA) Optics and Photonics Congress conference on "A Light in Digital Darkness: Optical Wireless Communications to Connect the Unconnected," in California, July 2019. The digital divide exists for the people without even any reliable Internet connection. High-altitude platforms (HAP)/UAVs and satellite constellation (GEO, LEO, CubeSats) backhauls to connect with the ground stations seem to be the most viable solution for establishing worldwide connectivity in remote locations. In this respect, laser satellite with site diversity and future optical wireless mesh network seems to be the best practical solution.

Creating and finally writing this book started stemming from three reasons: (1) tremendous growth of free-space optics (FSO) as evidenced by the applications of optical wireless communications (OWCs) to the information technology specifically to address the demand for a growing increase in data traffic due to widespread use of smart devices anytime, anywhere; (2) recent success of some demonstrations of laser satellite communication; and (3) proposed developments and deployments of mega-constellations of satellites (note: not necessarily using laser/optics-based satellites, yet!). However, the deployment of practical laser satellites servicing the network providers for providing high-speed wireless broadband Internet and telecommunications on a regular base does not exist. On the other hand, all-optical technology concepts to achieve these goals establishing laser satellites are available today. This book addresses this missing link of "how-to" design, develop, and analyze communication systems to achieve this goal of laser satellite-based connectivity and communication using satellite, aerial, terrestrial, and underwater optical terminals.

A number of recently published and edited books by the author, namely (1) Free-Space Laser Communications: Principles and Advances; (2) Advanced Free-Space Optics (FSO): A Systems Approach, both published by Springer, NY; (3) Optical Wireless Communications for Broadband Global Internet Connectivity, Elsevier, Amsterdam; and (4) Principles and Applications of Free Space Optical Communications, Edited Book, published by the Institution of Engineering and Technology (IET), London, already have paved the way to address the immediate next challenge and today's urgent need for providing practical solutions for global Internet connectivity based on FSO and OWC technologies. However, previously published books have not specifically discussed the laser satellite communication to include both a single and a constellation of satellites in various orbits (GEO, MEO, LEO) connecting with each other (e.g., inter-satellite links (ISLs)) and to optical ground stations (OGSs). This is the main reason to write this book. The book provides fundamentals of all-optical communications by 3D space optical networks and a comprehensive, unified tutorial to the understanding of fundamental issues of space optical networks to establish communications through the Earth's atmosphere and terrestrial and space environments in various orbits of the satellites. The book discusses the various communication architectures and scenarios for establishing global all-optical connectivity and communications. There is no other book available today to cover all the topics discussed in this book, combining the fundamentals and recent advanced technology in a concise but comprehensive manner. This book discusses the technologies for FSO and OWC as a practical solution for creating three-dimensional global broadband communication grid integrating satellite, aerial, terrestrial, and underwater laser/optical terminals to establish Internet access and communication anytime, anywhere, offering bandwidths and data rates far beyond the possible radio-frequency range and capacity. The book is therefore very timely because it can help researchers, students, faculty, and scientists in various laboratories and institutes in generating new and innovative ideas to design practical laser satellite-based communication systems.

The reader will also find it useful reading the theory to back up the all-optical communication systems with a distinctive focus on the impact of system performance and the broad range of applications for today's communication marketing needs. The book can serve as a textbook for graduate-level course and a reference book for professional engineers/scientists interested and involved in designing practical communication systems and for the researchers interested in this field.

In the last chapter, the book addresses an important subject of the impact and issues of existing and planned large satellite constellations for communications. The book lists and summarizes how the satellite constellations impact various space risks.

The outline and organization of the book were chosen such that the subject matter becomes more specialized with preceding chapters without losing the continuity. The book has nine chapters, each chapter having complete entities in and of themselves.

Chapter 1 describes how satellite wireless communications is an essential link for a connected world. Fundamental concepts of satellite communications with emphasis on the basic metrics for satellite communications and link performance for communications between a satellite and a ground station as well as for inter-satellite links (ISLs) are explained. Lead-ahead angles for establishing a link between a ground station and a satellite, between two low Earth orbit (LEO) satellites, and between two geostationary orbit (GEO) satellites are presented. Doppler shifts for satellite communication systems and potential compensation techniques relevant to GEO, LEO, and lower orbit satellites are addressed, which are essential in designing and developing satellite communication systems. Various satellite orbits and different types of satellites at different orbits are compared. Single-satellite all-optical technology concept for providing global broadband connectivity is introduced. Finally, a communication architecture (three-layer networks) to integrate satellite communications into terrestrial and other satellite networks using laser/optical (FSO) links is discussed, and the concept of satellite constellation networks is introduced.

Chapter 2 discusses the fundamental physics of space optical propagation applicable to integrated space/aerial, terrestrial, and underwater communication links. Slant-path free-space optical (FSO) communication links between an optical ground station (OGS) and a satellite at various orbits and altitudes including GEO, LEO, HAP, small, or even microsatellites are analyzed with appropriate accepted optical channel models and atmospheric propagation parameters. Various FSO transmitting beam shapes to optimize long-distance data transmission are also discussed. FSO communication technologies for 5G networks and future 6G evolution are explained. Underwater free-space optical (uFSO) communications for potential global connectivity at Gbit/s data rate is also presented.

Chapter 3 presents some of the fundamentals of laser satellite communications such as systems and technologies necessary to design laser satellite communication systems. Various types of GEO and LEO for various optical links are explained, which are necessary to establish all-optical global communications and connectivity. Essential results of optical propagation theory relevant to laser satellite communication are analyzed with emphasis on uplink and downlink wave models. A new technology concept for remote sensing of scintillation index is described. Laser communication technologies and link design fundamentals are also discussed. The relation between average signal-to-noise ratio (SNR) and average bit error rate (BER) necessary for establishing laser satellite links is explained with an example of high-altitude platform (HAP)-to-HAP optical link. This chapter provides a background and a solid foundation to move on to the understanding of laser satellite constellation in the next few chapters.

Chapter 4 discusses the development of high-speed (10s of Gbit/s or more) optical networks by integrating space (conventional satellite, small and microsatellites, high-altitude platforms (HAPs), UAVs, balloons, airborne), terrestrial, and underwater platforms. The global communication connectivity and Internet access will thus be possible if the airborne satellites and the fixed terrestrial terminals can all be integrated with appropriate optical devices and laser/optical transceivers. Critical issues and challenges of integrating space, terrestrial, and underwater communication links for designing FSO communication system are addressed. Some of the issues discussed are atmospheric turbulence effects on FSO communication system performance, underwater optical communications (uFSO) system performance, and an example of laser communication from a satellite to submarine using blue-green laser wavelength.

Chapter 5 discusses current advanced research relevant to "all-optical" communications and connectivity and future directions to make the concept of laser satellite constellation for achieving very-high-data-rate, high-capacity communication systems successful. This chapter presents various technology developments, research challenges, and advances in FSO communication system as well as discusses research challenges to be addressed and solved to overcome the limitations of existing systems and emerging new technologies. The recent applications related to FSO wireless communication systems also include space (aerial)-based and underwater platforms, 5G/6G networks, smart cars, robots, optical links for aiding health-related services, and artificial intelligence (AI)-based machine-to-machine (M2M) devices for adaptive optics (AO).

Chapter 6 presents the design concept technology leading to high-bandwidth, high-data-rate global Internet connectivity and seamless communications using all-optical technology based on integrating laser/optical communication constellation satellites, UAVs, balloons, and air terminals. The analysis is discussed to determine the optimum number of satellites in a given orbit for best communication performance and different communication links for the constellation satellites. Various types of satellites involved in communication constellation, conventional communication satellites, small satellites, microsatellites, CubeSats, and picosatellites: specific orbits: optimum number of satellites for given orbital heights and data rates and minimizing of latency in the network, are discussed. Applications of free-space optical (FSO) communications to design the basic architecture of space optical communication networks are described. Constellation of satellites for laser space communication network using GEO, MEO, and LEO satellites are addressed for both inter-satellite links (ISLs) on the same orbit and interorbital links emphasizing relay satellite communication network using constellation. Finally, constellation designs for LEO mega-constellations for integrated satellite and terrestrial networks to establish global communications and connectivity are presented.

Chapter 7 introduces the developments of concept technologies needed for establishing global broadband communication and connectivity using satellite constellations at different orbits. Each satellite belonging to a constellation will be equipped with laser/optical transceivers for transferring data communication information between them as well as from/to the constellation of optical ground stations (OGSs). This chapter discusses the most recent free-space optical (FSO) communication technology advances to achieve all-optical high-capacity communication systems for seamless global communication system performance. This chapter presents the concepts of optical satellite space networks relevant to constellation design and covers the establishment of satellite-aided Internet. Some of the device technologies include laser beam steering technology with no moving parts and MEMS-based fast steering mirror specifically useful for CubeSat constellation. Inter-satellite communication system is also addressed for future development of constellation of satellites. Finally, satellite-based global quantum communications and integrated space networks are also discussed. Challenges and progresses for implementing quantum key distribution (QKD) over long distance across free-space channels are also specifically addressed in this chapter. Recent developments in implementing QKD for LEO-to-ground link as well as for inter-satellite links (ISLs) in the presence of atmospheric turbulence are discussed and explained.

Chapter 8 discusses the important features of developing high-data-rate space communication networks using constellation of laser satellites, high-altitude platforms (HAPs), UAVs, and aerial, terrestrial, and underwater terminals connected with FSO laser/optical links. This chapter introduces relaying techniques as an effective method to extend coverage, which is essential in developing and designing reliable and cost-effective communication satellite constellation. All-optical amplify-and-forward (AF) and decode-and-forward (DF) relaying schemes relevant to FSO laser links between ground and GEO-LEO satellites, between satellites in a constellation, or between HAP and UAV-to-LEO scenarios are explained. Recent demonstrations and technology developments for FSO laser satellite communications relevant to both a single and a constellation of laser satellites are discussed. Some of the current projects and future missions of establishing laser communications in LEO and CubeSats are also mentioned.

In Chapter 9, summary, conclusions, and future directions are discussed, which include a summary of this book's major contributions and an important topic of impacts and issues of existing and planned large satellite constellation for communication. Some of the conclusions and future directions of the subjects covered in this book are discussed. Some of the challenges and future research of all-optical technologies using satellite laser terminals for establishing global connectivity and communication seamlessly with any space network are also mentioned in this chapter. Specifically, the types of FSO or laser ISLs and capabilities of terminals for future laser ISLs are addressed.

I want to thank Charles Glaser, Editorial Director, and Mary James, Executive Editor, at Springer Nature, New York, for initiating this book project and providing advice, comments, and inspiration. I also want to thank Arun Pandian KJ, Project Coordinator (Books) for Springer Nature, for efficiently organizing my book production process.

Over the years, I have benefited from technical informal discussions with many colleagues in the academic and research institutes in the areas of free-space laser/optical communications. Technical discussions and feedbacks from the participants, students, and researchers attending my teaching, invited lectures, and workshops were very helpful in emphasizing some of the topics of new concept technologies of current interests and needs. Some of these include IISc (Bengaluru), NITK (Karnataka), and Kings College of Engineering (Tamil Nadu) in India. I want to thank especially IEEE Photonics Society and Photonics Branch (India), Springer (IEEE EDS Kolkata Chapter), IEEE Communication Society and Optical Society of America (OSA), and South American Colloquium on Visible Light Communications (SACVLC 2021, Brazil, as International Advisory Committee member). I would also like to express my sincere thanks to Dr. Monish Chatterjee (University of Dayton), Dr. William Brown (Colorado State University Pueblo), and Dr. Timothy Brothers (Georgia Institute of Technology) for valuable technical discussions and suggestions. Discussions with Dr. Mikhail Vorontsov (University of Dayton) on remote monitoring of atmospheric turbulence, specifically the topology for deep machine learning, were beneficial.

I would like to express my sincere thanks to Dr. Morio Toyoshima, Director of Space Communications Laboratory, National Institute of Information and Communications Technology (NICT), Tokyo, Japan, for initial discussions at the early stage of preparing the book and providing me NICT's relevant research efforts. I want to thank Dr. Alberto Carrasco-Casado also of NICT's Space Communication Systems Laboratory and Dr. Morio Toyoshima of NICT for organizing an invited seminar and giving me an opportunity to visit NICT laser/optics laboratory and for valuable technical discussions.

Finally, I wish to thank my wife Gargi, my daughter Sharmistha, and her husband Nehemiah Benjamin for their support and inspiration throughout the preparation of the book and in completing my book under extreme adverse conditions during Covid-19/omicron situations. I also want to thank my sister, Dr. (Mrs.) Nilima Sarkar (retired from Boston Biomedical Research Institute, BBRI, and Harvard Medical School), for her encouragements. Arun K. Majumdar, San Diego, California.

San Diego, CA, USA   Arun K. Majumdar

Contents

Chapter 1
Basics of Satellite Wireless Communications: Single Satellite and a Constellation of Satellites

1.1 Introduction

Satellites play a crucial role to improve today's digital world dealing with almost every aspect of lives, from entertainment, business, banking, and agriculture to transportation and healthcare industries, and they all depend upon satellite communication technology. The United Nation's number of recent innovative progresses such as saving lives in emergencies and protecting environment as well as to connect the unconnected (thus to bridge the digital divide by bringing reliable Internet to those without it) are all examples where satellite wireless communication is essential.

1.1.1 Satellite Communications: An Essential Link for a Connected World

Satellite technologies are very diverse, but they all rely on one core element: the availability of wireless frequencies (radio frequencies and/or laser/optical frequencies) that can be operated reliably and be cost effective, safe, and affordable.

1.1.2 Motivation for Writing This Book: Why This Book?

The motivation for developing the concept of optical wireless communications for achieving global Internet connectivity using laser/optical communication satellites stems from the author's participation as an invited panelist at the Optical Society of America (OSA) Advanced Photonics Congress Symposium in San Francisco in 2019. The topic of "*A Light in Digital Darkness: Optical Wireless Communications*

© Springer Nature Switzerland AG 2022
A. K. Majumdar, *Laser Communication with Constellation Satellites, UAVs, HAPs and Balloons*, https://doi.org/10.1007/978-3-031-03972-0_1

to Connect the Unconnected" was relevant to the author's recently published book on optical wireless communications (OWC) for global Internet connectivity published by Elsevier in 2018 [1]. Some aspects of the OWC are also addressed in the author's recent online course, *"Optical Wireless Communications: Recent Application in Terrestrial, Space, Satellite and Underwater"* [13], sponsored by the IEEE Communication Society (ComSoc) Education & Training, in 2020 and 2021. The importance of developing satellite laser/optical communications concept technology to cover remote locations at high data rate is evident. Starting from a single-satellite communication, the technology is moving very fast to include a number of satellites, which is *constellation of satellites*. However, the OWC design for a single-satellite communication has to be modified to deal with a constellation of satellites specially with respect to different communication parameters such as different channels for different satellites, seamless handover issues, satellite orbits and altitudes, and latency problems and maintain communication continuity just to name a few. Some of the complex issues are (1) hybrid mesh global networks by bridging the divide between satellite-based and terrestrial networks, all with optical communication link; (2) concept technology of establishing potential global FSO networks from home to constellation satellites and from constellation satellites back to home in remote locations; and (3) the conceptual topology of integrated optical wireless constellation satellites, terrestrial, and home networks. All of these problems and possible concept technology will be covered in various chapters of this book, which hopefully provide the readers a comprehensive and thorough understanding of fundamentals and tools necessary to design laser/optical communication constellation satellite systems.

1.2   Fundamental Concepts of Satellite Communication

This section will discuss the fundamentals of communication satellite and will explain the parameters needed to understand first the operation of a single satellite, e.g., orbits, communication coverages, latency, handover, and ground station. The ultimate goal of this book is to develop the technology concepts of satellite constellation, which involve many satellites in various orbits to provide wireless communications *anywhere*, *anytime* including very remote locations worldwide. First of all, the goal of any type of communication is to send and transmit an information from one end and to receive the information at the other end via a communication link, which can be a wire, RF and microwave, fiber optics, or a free space like air. The received information must be undistorted and at the same speed or rate at both sending and receiving ends. Obviously, depending on the frequency, the signal sent can fall under the categories of radio frequency (RF) or optical wavelength. *Satellite communication* is the type of communication which takes place between any two (or more) Earth stations through a satellite. The communication involves transmitting, receiving, and processing of information such as voice, video, data, or any other information. Two other terms related to satellite communication are (1) a *repeater*, which can be a circuit (including processing) that increases the received signal and then transmits it, and (2) a *transponder*, which changes the frequency band of the transmitted signal from the received one (a repeater works as a transponder).

Why satellite communication?

The main function of a satellite is to receive signal (voice, image, data, or information of any kind) from Earth and transmit the same signal to a broad area (different locations) of Earth as well as to communicate with each other using inter-satellite links. The satellite can thus be used for relaying signals to a broad area of the Earth and therefore can provide communication in remote areas where other means of communication would be difficult to provide otherwise. Satellite communication system typically consists of space and ground segments as well as satellite control center (tracking, telemetry, and command station). The efficient use of satellite transponder can be a series of interconnected units forming the communications channel between the transmitting and receiving antennas (RF or optical).

There are many advantages of a satellite communication compared to the communication between two fixed points. There are also some disadvantages or drawbacks associated with a satellite communication. Some of them are described below.

Advantages of satellite communication:

Some of the advantages of satellite communication include the following:

- It can cover a wide area of the Earth, and also remote locations, which otherwise is either impossible or impractical. A number of variety of applications are possible such as weather forecasting, radio/TV broadcasting, global mobile communication including social networks and streaming videos, and navigation in emergency situations.
- It can be used for mobile and wireless communication applications.
- It can be incorporated with terrestrial line-of-sight (LOS) communication.
- It can manage easily the ground station sites.
- Satellite system can be interfaced with existing Internet infrastructure to obtain Internet services even in remote locations on the Earth. Internet satellite links are very flexible and can take advantage of communication networks in terrestrial, aerial, and space scenarios.

Disadvantages:

- Development cost is high in both designing and manufacturing.
- Debris from the satellite require attention. For example, as of January 2019, more than 128 million bits of debris smaller than 1 cm, about 900,000 pieces of debris 1–10 cm, and around 34,000 pieces of larger than 10 cm were estimated to be in orbit around the Earth. The debris can create "space pollution" (https:// en.wikipedia.org/wiki/Space_debris). This is of serious concern now and for the future.
- Monitoring and control are required to make sure that satellite is in the orbit.
- Lifetime is of the order of 12–15 years. Another launch needs to be planned a few years ahead of time.

- For designing a constellation of communication satellites, it requires a large number of satellites and very fast satellite-to-satellite handover in order to maintain visibility and continuous communication. This becomes a complex design issue.

The goal of this book is to develop the concept technologies required for the design of laser/optical links in space and in the atmosphere covering indoor, terrestrial, aerial, and space terminals. The chapter explains and discusses the details of (1) atmospheric optical/laser propagation and various mitigation techniques; (2) end-to-end laser/optical communication systems and subsystems including optimum modulation and detection techniques to send and receive very high data rate (thus taking advantage of the inherent capability of optical wavelength for achieving very high data rate and high capacity communication); and (3) evaluating laser/optical communication system parameters (SNR and BER) for all communication channels. This chapter discusses the basics and foundations for choosing satellite optical communication and various communication architectures.

Broadband communication delivery using satellites addresses the most important issue of "accessibility" and therefore is also best suited for areas underserved or unserved by terrestrial networks. There are many excellent books on satellite communications and the readers are referred to Refs. [5, 6], and therefore they will not be repeated here in details. But most of the books discuss the satellite communications using RF frequencies where the technologies have been around for many years. However, as mentioned previously, this book deals entirely with laser/optical communications as applied to satellite communication. This chapter describes some of the basic features of satellite communication, which apply to both RF frequencies and laser/optical wavelength for communication.

1.2.1   Basic Metrics for Satellite Communications: Brief Summary

The ultimate purpose of communications is to send intended information/message from a sender's source to the destination source where the information/message is received almost instantly (i.e., no or as small as possible latency) and vice versa. In addition, the data rates for transmission and reception should be at very high speed to satisfy today's demands for enormous amount of video and data exchanges. Satellites are also essential in mobile communications where, for example, billions of cell phone users need to be connected by Internets for audio, video images, and data exchanges among the users worldwide. A simple configuration of a satellite communication includes both ground segment and space segment (controlled by tracking, telemetry, and command stations). The space segment can also include inter-satellite links connecting the satellites. The *gateway, user, and hub stations* from the ground segment communicate with the space segment via both uplink and downlink communication links to establish two-way communication. Both radio or optical modulated carrier can be designed to establish such communication links between transmitting equipment and receiving equipment. For satellite communication scenarios, the communication channel is the space and terrestrial environment which includes atmospheric turbulence and scattering medium. For some special applications such as for satellite-to-submarine communications, the communication channels will include space, atmospheric turbulence, and ocean water medium.

Note that a single satellite can act like a network node when multiple stations transmit their carriers simultaneously to that satellite using multiple access scheme. If the satellite is equipped with *multiple-beam* RF or optical antennas, the process of routing of the carriers allows to receive a number of inputs and to send to different satellite channels as outputs with onboard processing and switching method. This way, the multiple beam antenna can send back the communication links to a number of selected coverage areas, which is defined by the *footprint* on the ground. The size of the footprint is determined by the carrier frequency (wavelength), the transmitting aperture diameter, and the satellite altitude. The field of view of the satellite for each individual antenna of the multi-beam antenna should be within the desired transmitting coverage area on the ground. Since the satellite is moving on the fixed orbit (either circular or elliptic), it should cover various geographical locations as it moves on the orbit. Description of various satellite orbits is given later in this section.

1.2.2   Link Performance Basics

There are basically two general configurations for satellite links to be considered where the satellite can move in any type of orbit:

1. Communication link between a satellite and a ground station: the communication channel is the space, atmospheric channel extending for a few kilometers above the Earth's surface.
2. Inter-satellite links between satellites: three types of inter-satellite links are the most considered in practice, namely GEO-to-LEO links between geostationary (GEO) satellites and low Earth orbit (LEO) satellites, GEO-to-GEO links between geostationary satellites, and LEO-LEO links between low Earth orbit satellites.

This book mainly aims to develop all-optical concept technology for broadband global communication and Internet connectivity. Therefore, only free-space optical (FSO) communication links and their performance will be addressed in this section. FSO communication offers a number of advantages such as very-small-size laser transceivers, small weight and low power consumption, capability of efficient large volume of data handling, high-speed data relaying, and implementation of complex networking software. Laser satellite communication has the potential of establishing all-optical global broadband communications and Internet connectivity *anytime, anywhere*.

1.2.2.1   Communication Link Between a Satellite and a Ground Station

Atmospheric effects such as atmospheric turbulence and scattering medium can ultimately limit the achievable maximum data rate. Communication link analysis provides a step-by-step procedure to calculate the received optical power as a function of transmitted laser power, communication range, atmospheric turbulence parameters, modulation schemes, and coding schemes. The author's previous books [1, 4, 14] have already discussed the detailed analysis and descriptions of the communication link analysis procedures with examples which are applicable to both uplink and downlink scenarios for satellite to ground. The link analysis discussed is applicable to analyze communication link between a satellite and a ground station. It is important to note that the atmospheric turbulence effects for the uplink laser beam are quite different than the downlink beam which needs to be considered for accurate design of laser satellite communications systems. Some of the parameters to be considered for laser satellite communication system design are latency between the sender and the destination sources, Doppler frequency shifts due to the motion of the satellite (influencing digital interface characteristics between satellite and terrestrial networks), and handover issues.

1.2.2.2   Inter-satellite Links

The laser-transmitting wavelength in a satellite has to be chosen to minimize interference between inter-satellite links and terrestrial systems, and therefore depends on the transmission characteristics of the communication channel like free space and any part of the atmosphere. Typical laser wavelengths in the regions of 0.8–0.9 and 1.06–1.55 μm are generally chosen. Some of the factors to be considered for optical links are the following:

Establishing a Link

The telescope size for optical link is typically of the order of 30 cm, and the transmitting beam angle is typically 3–5 μrad which obviously has negligible interference between systems. However, because of such a narrow beamwidth, it is essential to design an advanced pointing device which is a complex difficult problem. The three stages of establishing the optical link are acquisition, pointing, and tracking (ATP), and then only communication can be established to exchange information between the two ends.

Lead-Ahead (or Point-Ahead) Angle

The lead-ahead angle, or sometimes referred to as point-ahead angle, arises when the relative motion of the satellite either relative to a ground-based transmitter or between two satellites in case of inter-satellite link is considered. Some of the scenarios are discussed below.

A Ground Station and a Satellite

When the observation points where the laser transmitter is located at the optical ground station (OGS), the ground-based transmitter is then pointed at the position where the satellite will be when the beam arrives (see Fig. 1.1). This requires a lead-ahead angle $\theta_P \cong 2V_T/c$, where V_T is the speed of the satellite perpendicular to the line of sight and c is the speed of light.

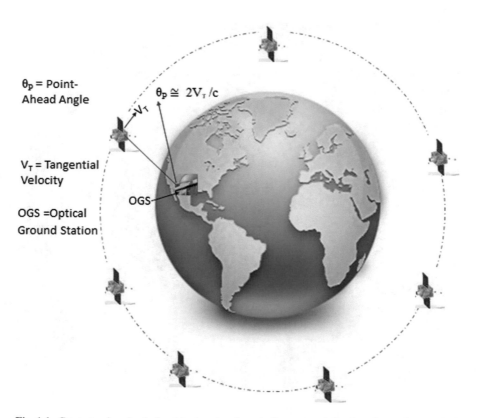

Fig. 1.1 Geometry for a lead-ahead (point-ahead) angle for a transmitting laser beam from optical ground station (OGS) to an orbiting satellite

Two LEO Satellites

For the case of inter-satellite link (ISL) between two satellites, the concept of lead-ahead angle can be understood as follows. There are two satellites, the first one at A and the second one moving along the trajectory B, C, and D. In order to establish an ISL between the two satellites, for example between the first satellite at A and the second satellite at point C, it is important to take into account the relative motions of the two satellites. The first satellite at A and the second satellite at C are moving with velocity vectors V_{T1} and V_{T2}, respectively, with components orthogonal to the line joining the two satellites at time t as depicted in Fig. 1.2. The lead-ahead angle between two satellites at A and C with velocity vector components V_{T1} and V_{T2} in a plane perpendicular to the line joining the two satellites at time t and t_{tp} is given by [5]

$$\beta = 2\left|V_{T1} - V_{T2}\right| / c \left(\text{rad}\right) \tag{1.1}$$

where β is the lead-ahead angle which is actually the angle between the second satellite at B at time $t - t_p$ and the second satellite at D at time $t + t_p$.

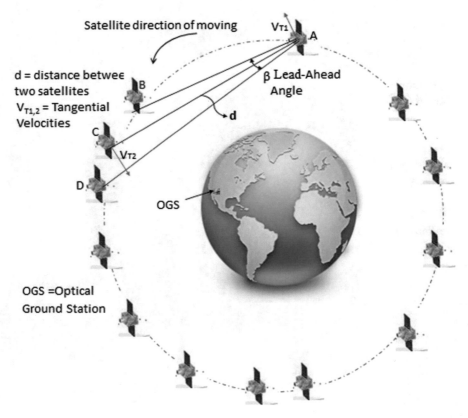

Fig. 1.2 An artist's concept to define the lead-ahead (point-ahead) angle in establishing inter-satellite link (ISL) communication

$t_p = d/c$ = the propagation time of a photon from satellite at A to satellite at C, d = distance between the two satellites, and c = velocity of light ($c = 3 \times 10^8$ m/s).

Two GEO Satellites Separated by an Angle

For ISL between two GEO satellites on the same orbit, the lead-ahead angle for two GEO satellites separated by an angle α is given by [5]

$$\beta = 2|V_{T1} - V_{T2}|/c = 4\omega(R_o + R_E)\sin\left(\frac{\alpha}{2}\right)/c\,(\text{rad}) \qquad (1.2)$$

where

ω = angular velocity of a GEO satellite = 7.293×10^{-5} rad/s
R_o = altitude of a GEO satellite = 35,786 km
R_E = radius of the Earth = 6378 km

A GEO Satellite and a LEO Satellite with Circular Orbit

Lead-ahead angle for a GEO satellite and a LEO satellite with circular orbit and with velocity vectors V_{S1} and V_{S2} can be written as [5]

$$\beta = \frac{2|V_{T1} - V_{T2}|}{c} = (2/c)\left\{|V_{S1}|^2 + |V_{S2}|^2 - 2|V_{S1}||V_{S2}|\cos i\right\}^{1/2}\,(\text{rad}) \qquad (1.3)$$

where i = the LEO satellite inclination

$$|V_{S1}| = \omega_{GEO}(R_o + R_E) = 3075\,\text{m}/\text{s} \qquad (1.4)$$

$$|V_{S2}| = \omega_{LEO}(h + R_E)$$

h = LEO satellite altitude.

1.2.2.3   Doppler Shift for Satellite Communication Systems and Compensation Techniques

Doppler shift occurs because of the relative motion between the satellite and the mobile terminals which can be either located on the ground or in space. The higher the satellite altitude from the ground is, the slower is the motion around orbital position to a point on the Earth. For example, LEO satellites at lower altitude than the GEO satellites experience a higher relative motion and therefore experience higher Doppler effects. The small satellites such as CubeSats or microsatellites located in much lower altitudes than the LEO satellites therefore experience the highest

Doppler shifts. It is essential to develop some compensation techniques for the Doppler shifts, especially for the satellites orbiting at altitudes close to the ground.

The Doppler shift is a change in frequency when a source of waves is moving relative to an observer or vice versa resulting in a change in frequency in relation to the observer. For an observer approaching the destination, the frequency increases, whereas if the source is moving away from the destination the frequency decreases. The observed frequency due to the relative motion between the satellite and the terminal is given by [15]

$$f = \left(1 + \frac{\Delta v}{c}\right) f_0 \qquad (1.5)$$

where $\Delta v = v_r - v_s$ is the relative motion between the satellite and the terminal, v_r is the velocity of the receiver, v_s is the velocity of the satellite, c is the speed of the light 3×10^8 m/s, and f_0 is the carrier frequency of the source. The Doppler shift is defined as the change in the carrier frequency and the observed frequency, $\Delta f = f - f_0$, which is therefore given by

$$\Delta f = \frac{\Delta v}{c} f_0 \qquad (1.6)$$

For LEO circular orbits, the angular velocity can be written as approximately [15]

$$\omega_F \cong \omega_S - \omega_E \cos(i) \qquad (1.7)$$

where ω_S is the angular velocity of the satellite and ω_E is the angular velocity of the Earth and is equal to $2\pi/(24 \times 3600)$. In the above equation, i is the inclination of the orbit.

Doppler equations for LEO satellites are discussed in detail in the research works [15, 16]. For LEO satellite communication, the Doppler shift is zero when the satellites are in the maximum elevation angle, which is at its closest position to the ground terminal equipped with transceiver, and the shift is larger at a lower elevation angle. Some of the following Doppler shift compensation strategies are discussed [15]:

– *Doppler shift compensation at the mobile terminal* requires the exact position of the satellite and the position of the terminal.
– *Doppler shift compensation at the satellite* requires the satellite coverage area for a static ground cell and the position of the center of the cell. The satellite's position is also needed.

The author also discussed the Doppler shift in different points of the cell and the Doppler shift as the function of the height of the orbit. Note that the Doppler shift actually degrades the communication system performance such as degrading the bit error rate (BER) for a given signal-to-noise ratio (SNR) when compared to the ideal one with zero Doppler shift. *It is therefore essential to accurately design the satellite-moving terminal system so that the Doppler shift is minimum.*

1.2.3 Description of Satellite Orbits

In this section, the orbits of various communication satellites will be discussed. Orbits of the typical satellites such as GEO, MEO, and LEO satellites will be described followed by the orbits of the recently developing satellites such as small satellites. All types of satellites have specific features and purposes for establishing satellite communication to satisfy different requirements. Some of the factors for the choice of a typical satellite communication system configuration and satellite orbit are the following:

- Geographical coverage requirement
- Service types (IP broadband, mobile satellite, emergency communication, fixed satellite)
- Look angle of the satellite from a ground station
- Spectrum availability and resource for the chosen orbit
- Cost, reliability, and lifetime of satellite network

Geostationary satellite (GEO): located at a height of about 36,000 km above the Earth; remains approximately fixed relative to the Earth; goes around once every 23 h and 56 min; maintains continuous positioning above the Earth's subsatellite point on the equator; and uplink/downlink transmission delay is about 250 ms which can affect when very high data rate of communication from the ground station is needed.

Medium Earth orbit satellite (MEO): located at altitudes ranging from 8000 to 15,000 km above the Earth; requires multiple satellites such as a constellation of about 10–15 satellites to maintain constant coverage of the Earth; reduced latency and improved look angles to ground stations; shorter lifetime than GEOs; and requires complex ground station for receiving and tracking satellite signals.

Low Earth orbit satellite (LEO): located at a height of about 1000 km above the Earth; free-space losses are smaller than GEOs and MEOs since these satellites are nearer to the Earth; lowest latency; higher received communication signa; better look angles; shorter lifetimes in orbit; requires a large number of satellites in the constellation to provide constant Earth coverage.

Figure 1.3 shows a sketch of the satellite orbits for GEO, MEO, and LEO showing the respective latencies in milliseconds.

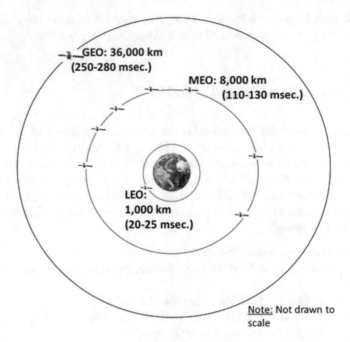

Fig. 1.3 The orbits for GEO, MEO, and LEO satellites with respective latencies in milliseconds. (Note: Not drawn to scale)

1.2.3.1   Small Satellites for Communication: The Next Space Revolution

In addition to the typical GEO, MEO, and LEO satellites located at their orbital altitudes discussed earlier, recently the size and weight of the satellites have dramatically impacted the changes of the future business and space explorations. Tens, hundreds, or even thousands of small satellites are being considered and developed which will make the next space revolution. Thousands of small communication satellites (constellations) should be able to give 100% of global coverage. The details of the design of constellation of small satellites with the big players for achieving optical wireless communications for broadband global Internet connectivity will be discussed in detail in a later chapter. One of the main objectives of using small satellites is obviously to bring ultrahigh-speed communication connectivity and Internet worldwide.

Small satellites and spacecrafts usually termed as "*smallsats*" are being designed to be useful for establishing global communications expanding access to space and can reduce the cost of new space missions considerably. NASA has already developed a Small Spacecraft Technology Program (SSTP) to demonstrate new small spacecraft technologies which are relevant of free-space OWC. SSTP funded the development and implementation of multiple launches for small satellite missions between 2017 and 2018, in particular for high-speed optical communications to increase downlink rates. Small spacecrafts defined by NASA apply to those with a mass of less than 180 kg and are classified as follows [6]:

- Minisatellite —100 kg or higher
- Microsatellite —10–100 kg
- Nanosatellite —1–10 kg
- Picosatellite —10 g–1 kg
- Femtosatellite —1–10 g

A *CubeSat* falls in the category of nanosatellites with a minimum dimension of 10 cm × 10 cm × 11 cm and has also been built in various larger sizes.

For comparison, the typical satellites are categorized as follows:

- Large satellite —1000 kg

- Medium-sized satellite —500–1000 kg

1.3 Space Laws Relevant to Satellite Communications

At this point, it is important to know about the space laws that govern space-related activities governed by both international and domestic agreements, rules, and principles [2]. The law includes space exploration, liability for damage, weapon use, rescue efforts, environmental preservation, information sharing, new technologies, and ethics [3]. The international treaties include 1967 "Outer Space Treaty," 1968 "Rescue Agreement," 1972 "Liability Convention," 1975 "Registration Convention," 1979 "Moon Treaty," and 1998 "International Space Station" (ISS) agreement. The law also encompasses national laws and different national space legislation passed by many countries and also geostationary orbit allocation ("Allocative Limitations") and covers issues such as environmental protection and ethics. Laws most relevant to satellite communication involve *spectrum management* for both international (International Telecommunications Union, known as ITU, to deal with countries) and national (to navigate the frameworks covering spectrum allocation) efforts to regulate the use of *electromagnetic spectrum*. Because of increased number of satellites, it is equally important to develop regulation regarding how increasing volume of data is being transmitted through space. For satellite communications, *cybersecurity* plays an important concern of interest to both national security and consumers. In addition, for communications satellites, it will be needed to ensure that orbital crowding does not reduce the ability to operate in crucial orbits (Space Situational Awareness, SSA).

1.4   FSO Communication Relevant to a Single (and Multiple) Satellite(s)

This section explains the urgent need for today's tremendous demand in this digital world, and how free-space optical communication (FSO) can provide practical solutions to meet these needs. Various categories of FSO are discussed, which include both indoor and outdoor FSO links. Outdoor FSO links will include terrestrial, aerial, and space terminals where hybrid optical networks will play a major role in establishing duplex communication for information exchanging.

1.4.1   Need to Address Demand for High-Speed Communications in the Digital World

This chapter discusses the demand for high-speed communication and addresses the problems, present and future, of the so-called *digital world* which is basically the availability and use of digital tools to communicate on the Internet, digital devices, smart devices, and other related technologies as well as human interactions that are suggested as part of the digital world. The rapid growth and expansion of the Internet, especially the World Wide Web (WWW) and the associated social media, are continually changing the media landscape. Some of the statistics include the following:

- *Ever-increasing number* of wireless broadband Internet, mobile phones, smart devices, social web, gaming, and video-centric applications. Over 80% of the world will be connected to the Internet by 2020.
- *Global Internet adoption and devices and connection.*
- *Internet users*: Nearly two-thirds of the global population will have Internet access by 2023: 5.3 billion total Internet users (66% of global population).
- *Devices and connections*: The number of devices connected to IP networks will be more than three times the global population by 2023—includes M2M connections and Internet-of-Things (IoT) by applications (connected home applications and connected car).
- *Mobility growth*: Over 70% of the global population will have mobile connectivity by 2023; 5G devices and connections will be over 10% of global mobile devices and connections; and smartphones will be the fastest growing mobile device.
- *Global network performance and data rates by 2023: Fixed* broadband speed will be ~110.4 Mbit/s, *mobile (cellular)* speeds 43.9 Mbit/s, 5G speeds ~575 Mbit/s, and Wi-Fi speeds from mobile devices ~92 Mbit/s (public hotspots will be ~628 million) (*Source: Cisco Annual Internet Report (2018–2023) White Paper, updated March 9, 2020*).

This chapter explores how best these new technologies can be utilized to enhance the quality of communication for the majority of the people. All the chapters in the book provide the only and most viable technology comprehensive solution using laser/optical communication with constellation satellites, UAVs, and balloons for global communications and Internet connectivity to build a real *all-optical global network*. The goal of the book is to present innovative design concepts for achieving very-high-speed (multi-Gbit/s) free-space optical (FSO) communications for terrestrial, space, satellite, and inter-satellite links, and then incorporating with worldwide constellation satellites. Today, people are constantly interested to have instant access to communicate in real time to other parts of the world using one of the most convenient devices like smartphones and other digital devices for the purpose of social networking (Facebook, WhatsApp, Twitter, etc.), entertainment (live broadcasting of sporting events, videos/video streaming, movies), business (commercial, marketing, financial information, banking, ATM), education (distance learning, online course, teaching in remote villages), global remote sensing (hydrology, ecology, meteorology, geology), medical (telemedicine, collaborative surgery help), and emergency situations in disaster (earthquake, fire, tsunami, humanitarian applications). Because of this recent enormous evolution of wireless communication applications, driven by the ever-increasing number of wireless broadband Internet, mobile phones, smart devices, social web, gaming, and video-based applications including video streaming, the number of end users is growing very fast more than 30–40% or even more per year.

1.4.2   Why RF Is Not a Viable Solution Compared to Optical Wireless Communication (OWC) for a Satellite Communication?

Some of the obvious advantages of laser over RF communication relevant to satellite communication are as follows: (1) high bandwidth of lasers versus microwaves: the frequency ranges of optical frequency from 10^{12} to 10^{16} Hz, data bandwidth can reach up to 2000 THz in case of using optical carrier for communication; for RF case, the RF range is comparatively lower by a factor of 10^5 [2]; (2) very narrow beam size which enables the transmitted power to be focused on a very narrow area, and very high directivity is attainable with small-size antenna using laser beam; (3) no license required; (4) no electromagnetic interference (EMI); (5) antenna size is considerably smaller (optical aperture lens acts as optical antenna); and (6) secure communication: RF beam spreads all over and the optical beam with very narrow beam width with LOS propagation offers much more secure communication. Some of the disadvantages of FSO communications include extreme stringent pointing requirements, and degradation of optical beam due to atmospheric effects such as turbulence and fog can limit the data transmission.

1.4.2.1   Comparison Between RF and Laser Footprints for GEO and LEO Orbit Satellites

For a laser beam with diffraction-limited optics, the minimum transmitted beam width that can be achieved is given by

$$\theta \cong 1.24\lambda / D \tag{1.8}$$

where λ is the laser wavelength and D is the aperture diameter. For a typical laser-com wavelength of $\lambda = 1.55$ μm and a transmitting diameter of 10 cm, the transmitted beam widths for GEO and LEO satellites are

$$\theta_{GEO,Laser} \cong 34.72 \,\mu rad$$

If the laser beam is focused from a GEO satellite directly along the equator, it will be a circle with a diameter (footprint) of $D_{GEO,Laser} \cong R_{GEO} \cdot \theta_{GEO,Laser}$ where R_{GEO} is the GEO altitude ($\cong$35,000 km). Therefore, $D_{GEO,Laser} \cong 784$ m (GEO laser footprint).

For RF signal at a typical frequency of 10 GHz ($\lambda = 3$ cm) and with a dish antenna of diameter of 1.0 m, the $D_{GEO,RF} \cong R_{GEO} \cdot \theta_{GEO,RF} = 2352$ km (GEO RF footprint).

For LEO, the same calculations yield the following results:

$D_{LEO,Laser} \cong 34.72$ m (LEO laser footprint)
$D_{LEO,RF} \cong R_{GEO} \cdot \theta_{LEO,RF} = 67.2$ km (LEO RF footprint)

Figure 1.4 shows the footprints of laser beam and RF beam for GEO and LEO orbit satellites. Smaller intercept areas allow more secure communication with higher received signal power.

RF-based wireless existing technology and systems *cannot* meet this requirement. Conventionally, RF spectrum is very congested and spectrum usage is restricted. Even in a highly populated indoor environment, the end users using RF-based wireless technologies experience lower data rates with poor quality of services because of RF spectrum congestions. The only viable solution is the free-space optical (FSO) communication where all end users will have access to optical communications with its tremendous bandwidth capability. Some of the features OWC can provide include license-free spectrum, much higher bandwidth and capacity applications, reduced power consumption (size, weight, and power, SWP), compactness, portability, ease of installation, and much improved channel security. With potential FSO solution, there is no existing *complete* optical solution for the end users in both locations to *worldwide/globally* connect broadband Internet, mobile phones, smart devices, social web, etc. "*All-optical*" technology combining satellite communications, free-space optics (FSO), and wireless LAN is the only potential and viable solution to establish worldwide broadband wireless access independent of terrestrial infrastructure, "*anywhere anytime*".

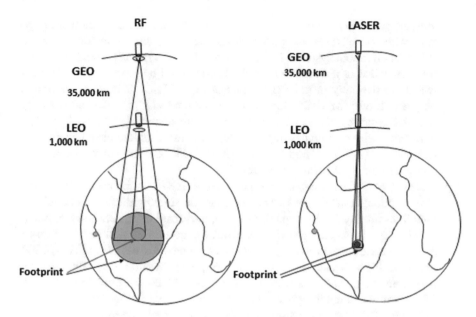

Fig. 1.4 The footprints of laser beam and RF beam for GEO and LEO orbit satellites. The footprints are approximately the ground intercepts on the Earth when the transmitting beam is focused directly down along the equator and depend on the altitude of the satellite and its pointing angle

This proposed book addresses an entirely different approach which is based on the constellation of laser communication satellites, which can solve the truly multi-Gbit/s or more global connectivity problem even in extreme remote locations, thus providing communications access *anytime anywhere*. FSO constellation satellite network is discussed and explained in this book to show how it can provide a high-bandwidth optical wireless network access to the end user since a constellation of satellites can cover almost any remote area on the Earth. Finally, combining FSO communications constellation satellites with underwater optical communication networks can also provide additional solution to cover almost any part of the global region.

1.4.3   Single-Satellite All-Optical Concept Technology for Global Broadband Connectivity

1.4.3.1   FSO Communication Systems: Single-Satellite FSO

Free-space optical wireless communication (FSOC) can be both indoor and outdoor OWC.

(a) *Indoor FSO Communication*: The indoor optical communication is a short-distance communication. The communication link performance factors for long-distance FSO communication apply equally to short-distance indoor systems with one obvious exception: atmospheric loss due to turbulence has no effect for all indoor systems. Indoor communication typically has a base station for each cell with a number of terminals connected by short-range FSO links, which are basically infrared (IR) links using LEDs or diode lasers at visible wavelength offering visible light communication (VLC) [1]. The communication performance parameters for indoor OWC are analyzed in terms of propagation model which computes impulse responses for an indoor channel with both line-of-sight (LOS) and non-line-of-sight (NLOS) path to evaluate the received power and the SNR for indoor system [1].

(b) *Outdoor FSO Communication*: The outdoor FSO communication links can have terminals outside the house (for example streetlights for VLC equipped with light fidelity (Li-Fi) capabilities [1]), terrestrial (e.g., between buildings), aerial (e.g., between ground and airlines), ground-to-satellite, satellite-to-ground, satellite-to-airborne platforms (unmanned air vehicles (UAVs)), balloons or high-altitude platforms (HAPs), and satellite-to-satellite. Recent developments in underwater optical communications can provide underwater FSO communication for global network connectivity [4]. Figure 1.5 depicts the different FSO communication scenarios as described above, which include indoor, terrestrial, airborne, satellite, and underwater platforms.

There are three basic types of Earth orbits relevant to communication satellite: communications architectures of different satellites as shown in Fig. 1.6. The figure depicts orbits for GEO, MEO, and LEO satellites with respective latencies in milliseconds. Each satellite's motion is mostly controlled by Earth's gravity as explained by Kepler's laws. The satellite's height, eccentricity, and inclination determine the satellite's path, and also the view of the Earth. Since the geostationary satellites, GEOs, are always over a single location, they are very useful for communication (phones, TVs, radio, and other video and data). The satellites in the MEOs move more quickly and are classified as the semi-synchronous (near-circular orbit, low eccentricity) orbit and the Molniya (highly eccentric) orbit types. Satellites in the low Earth orbit (LEO) have nearly circular orbit, and the inclination depends on the monitoring types of the satellite.

Recently, *small satellites* for various applications are receiving considerable attention, and a microsatellite is one of the types of small satellites which have mass in the range of 10–100 kg, and offers some advantages like low cost for development and launching, high operational flexibility, and potential *platforms* for *future laser communications*. In order to understand the capability of future potential laser communication between a microsatellite in LEO and a dedicated optical ground station (OGS), it is important to note that a microsatellite in LEO altitude is able to communicate at most several times per day. This is determined by the orbital geometry which includes the satellite's height, eccentricity, and inclination that determine the satellite's path, and also the view of the Earth as mentioned earlier. Furthermore, the duration of each communication window is only up to about 10 min.

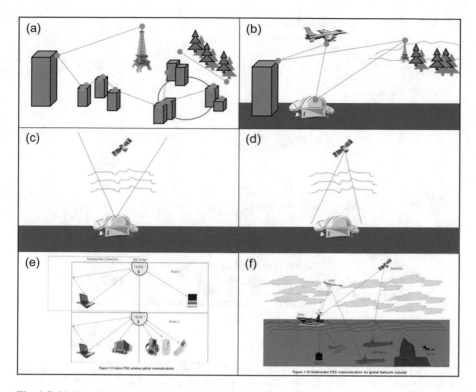

Fig. 1.5 Various FSO communication scenarios including indoor, terrestrial, aerial, satellite, and underwater communications. (*Reprinted with permission from Springer Nature and Copyright Clearance Center, 2021* [4])

Many of the LEO satellites have a nearly polar orbit (highly inclined orbit), and the satellite moves around the Earth from pole to pole which takes about 99 min to complete an orbit. For the one half of the orbit, the satellite views the daytime side of the Earth, whereas it crosses over to the nighttime side of the Earth at the pole. For a LEO communication satellite during staying on the orbit, the Earth underneath of the orbit also turns. This is important from communication point of view establishing LOS communication link also shifts due to the Earth's movement and communication will require very precise pointing and acquisition accuracy to exchange communication from the original ground station point on the Earth. To stay over one point on Earth (ground communication station), polar-orbiting satellites can stay on the *Sun-synchronous orbit* which allows the local time on the ground to be same whenever and wherever the satellite crosses the equator. A Sun-synchronous orbit crosses over the equator approximately at the same local time each day and night and also allows to keep the angle between the Sun and the Earth's surface relatively constant. The researchers illustrated [7, 10] three consecutive orbits of a Sun-synchronous satellite with an equatorial same crossing time. Note that the path a LEO satellite at a height of 100 km to stay in a Sun-synchronous

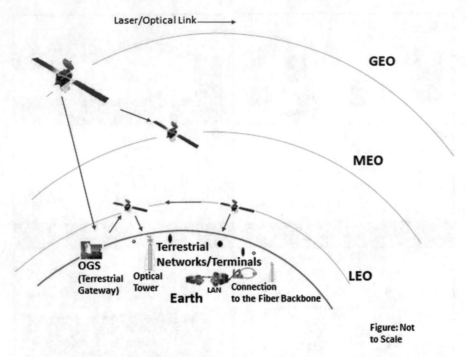

Fig. 1.6 A sketch of communication architecture (three-layer networks) to integrate satellite communications into terrestrial and other satellite networks using laser/optical (FSO) links. (Figure: Not to scale)

orbit is very narrow and will need an orbital inclination of 96°. Since the period for one orbit of a microsatellite is about 90 min so that the microsatellite cycles around the Earth about 15 times per day with the number of communication sessions is about 3–4 with the duration of each communication window is only up to 10 min [8] in order to maintain the direct communication to ground station. From Fig. 1.6, it is apparent that there is a LOS between the satellite and the ground station in the first orbit, but in the next orbit there is no LOS between the satellite and the ground station because of the rotation of the Earth on its axis [7, 8] which results in no communication. These are some of the drawbacks of a microsatellite to maintain long-time communication. The microsatellite with a data rate of 1 gigabit per second (1 Gbit/s) can transfer ~300 gigabytes (GB) to the ground in one day. To compare with GEO, if 5 Gbit/s data rate using laser communication is used for communication, the LEO satellite can transmit about 27 terabytes (TB) per day to the GEO satellite and the link availability is 100% [8, 9]. Also as mentioned in [8], 2552 TB of the user data can be transmitted from the microsatellite to GEO to the ground per year, and at 1 Gbit/s data rate, the amount of data which can be relayed to the ground is about 5440 GB/day. *Thus, GEO satellite can be used as a relay satellite to transfer data from a microsatellite at LEO to the optical ground station.*

The challenge of developing small satellite-based laser/optical communications systems is getting the data to the optical ground station (OGS) with a potential of multi-Gbit/s download speed. Communication performance between a satellite in an orbit flying within the range and a ground station depends on a few factors: (1) altitude of the satellite orbit, (2) latency (this is already determined by the altitude of the orbit), (3) number of times satellite cycles around the Earth per day (this also determines the number of communication sessions), (4) duration of each communication window (to maintain the direct communication to ground station), and (5) pointing and tracking of the laser beam. Some of these factors are discussed earlier in this section and are valid for a single satellite. Another alternative approach to solve the above problems is to develop a network of small LEO satellites to act as small optical relay satellites connected to the optical network. A proper design is therefore required to implement optimum optical downlink as well as optical cross-links for a small satellite. This is the same as if a constellation of satellites is designed and developed which will be discussed in detail in the other chapter.

1.4.3.2   A General Space-Based Laser/Optical Communication Platform Architecture

A space-based laser/optical communication platform architecture requires an infrastructure that can be used for various applications and requires a network of satellites in the equatorial plane, optical ground stations (OGSs), and end-user terminals that are interconnected. Therefore, different types of satellites and their design are needed to satisfy the requirements of the end users. In order to transmit data from satellites directly to ground stations, a direct line of sight between the satellite and the receiving ground station is needed, and this is only available for a few minutes per day for a small LEO satellite. More efficient architectures than existing current systems are required.

Figure 1.6 illustrates a sketch of communication architecture to integrate satellite communications into terrestrial and other satellite networks using laser/optical (FSO) links. The architecture encompasses satellites in different orbits and terrestrial terminals, which can be end-user nodes (thus can be a part of LAN and WAN networks). There are two ways the end users can be connected in this communication network: (1) Ground terminals communicate directly with the satellite in the LEO orbit. The limited capabilities of the ground terminals in terms of the duration of each communication window impose challenges for continuity of communication with a single satellite. (The problem can be practically solved by designing a constellation of such LEO satellites which will be discussed in a separate chapter.) (2) The second option is to make end-user communication nodes communicate through a terrestrial gateway/relay node where the constellation of satellites acts as the backhaul. This concept will also be explained in a separate chapter. Figure 1.6 also suggests that the hybrid architectures combining different GEO, MEO, and LEO satellites with *FSO* links will be the future solution. Small satellites in LEO orbits can thus overcome the short transmission times and issues with the wide coverage and other related communication performance capabilities of GEO satellites. This way, an extreme flexible global communication network can be developed using FSO links for achieving both the flexibility and large capacity.

A GEO satellite is at the altitude of 36,000 km and moves in the orbit with the same velocity of the Earth' rotation keeping the same relative position with a fixed ground station. In the relay system just mentioned, GEO satellite can be the relay terminal, and the relay communication system can work by first sending the communication data to the GEO, which can then transmit the data directly to the ground station. Of course, a number of factors such as latency and divergence of the laser beam reaching the ground terminal need to be considered via communication link equation and mitigation of atmospheric effects (including proper adaptive optics technique) in order to maximize the received SNR at the fixed optical ground terminal.

The footprint size from a transmitting satellite depends on the laser-transmitting beamwidth, wavelength, and altitude of the satellite and is depicted in Fig. 1.7, which sketches an artist's concept for transmitting laser beam widths for (1) downlink of GEO-to-ground station, (2) uplink from LEO to GEO, and (3) relay satellites within LEO orbit (drawn not to scale). As mentioned before, the importance of pointing of laser beam to achieve high data rates with moderate power by focusing the transmitted power into a narrow beam to direct to a receiver so that the receiver is reasonably centered in the beam profile. Data rate is also affected by the transmitted power, the size and efficiency of the collection optics and receiver, and the communication range. Beam divergence and range play an important role in collecting photons into the collector so that the photons can be converted to ultimately data bits. For example, the transmitter pointing accuracy (degrees) from a 4 W laser to a 10 cm diameter receiver for a typical LEO-to-ground link (~1000 km) for 1 Gbit/s data rate is about 0.013°, and for GEO-to-ground link (40,000 km) is about 0.0003° [11].

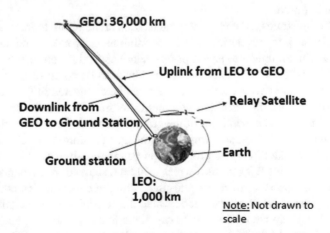

Fig. 1.7 Sketches an artist's concept for transmitting laser beam widths for (1) downlink of GEO-to-ground station, (2) uplink from LEO to GEO, and (3) relay satellites within LEO orbit. (Note: Drawn not to scale)

Therefore, a LEO transmitter with sufficient pointing accuracy for a GEO satellite (~40,000 km) should also have adequate pointing requirement for LEO at 1000 km downlink at the same data rate.

1.4.3.3   Small Satellites for Potential Laser/Optical Communication

Recently, there has been an increased interest in developing low-cost small satellites for various applications. One of the main factors for this is the miniaturization of fast optical components and devices, which are making it possible to fit in a small-size satellite to act like a conventional satellite. In other words, there is an increased utilization of COTS components to build a small satellite effectively. There is also a rise in the small satellite constellations and increased deployment of small satellites by commercial players. This section discusses about some of the parameters of small satellites, which are relevant to communication. The different categories of small satellites are already mentioned earlier. Nanosatellites are launched in low circular or elliptical orbits between 400 and 650 km and travel around 8 km/s. It takes about 90 min to orbit the Earth and completes between 14 and 16 orbits a day. From communication point of view, nanosatellites are useful in developing the Internet of Things (IoT) on a global scale to make the necessary global connections using communication networks.

A comparison of various small satellites is shown in Table 1.1 showing some of the parameters.

The calculated values of laser footprint for the small satellites using Eq. (1.1) as explained earlier are shown in Table 1.2 for two values of the laser-transmitting aperture diameters, D: $D = 10$ cm and $D = 5$ cm. This is very useful to determine the feasibility of potential laser/optical satellite communications for these types of small satellites.

The laser spot size on the Earth depends on the laser-transmitting aperture size, D, and the optimum aperture size will have to be designed from the other dimensions of the small satellite. But it is encouraging to explore the potential use of *laser-based small satellites* to provide high-data-rate communications and also incorporate the fast-developing *IoT* applications.

Table 1.1 A comparison of various small satellites [12]

	Mass (kg)	Altitude (km)	Orbit period (h)
Mini	100–500	1000–5000	2–3
Micro	10–100	500–2000	1.6–2
Nano	1–10	300–800	1.4–1.7
Pico	0.1–1	200–400	1.4–1.5
Femto	<1000 g	200–400	1.4–1.5

Table 1.2 A comparison of laser footprint for various small satellites

	Altitude (km)	Laser footprint for $D = 10$ cm (m)	Laser footprint for $D = 5$ cm (m)
Mini	1000–5000	34.72–173.6	69.44–347.2
Micro	500–2000	17.36–69.44	34.72–138.88
Nano	300–800	10.4–27.8	20.83–59.55
Pico	200–400	6.94–13.89	13.89–27.78
Femto	200–400	6.94–13.89	13.89–27.78

1.4.4   Communications Architecture of Satellites: Introduction to Satellite Constellation Networks

It is evident that satellite communications are essential for connecting the two users located in two different locations to exchange communication information. This is the first book to specifically address global laser/optical communications using *constellation satellites*, UAVs, HAPs, and balloons to satisfy the tremendous increasing demand for data exchange and information between end users worldwide including remote locations. This book is an essential reading for all optical communications, telecommunications, and system engineers, as well as technical managers in the aerospace industry and graduate students and researchers in academia and research laboratory. Later chapters in the book will discuss and explain the essential components for achieving all-optical concept technology that will require three things:

1. Very high communication capacity (of the orders of a few Tbit/s, depending on the orbital heights and the number of satellites).
2. Satellite-to-satellite (within the constellation) optical cross-links (with a speed of about 200 Gbit/s).
3. Satellite-to-ground optical bidirectional links (handling about 100 Gbit/s data capacity) The all-optical global connectivity scheme can also eventually include Internet traffic from underwater platforms like submarines to satellites.

Next few chapters in this book also discuss step-by-step methods to develop satellite backbone in order to interconnect a number of ground nodes clustered within a few SD-WAN (software-defined networking) in a wide area network (WAN) around the world in order to provide a fully meshed communication network. The challenges of laser transmission and reception are affected by weather and clouds, fog, and rain affecting a satellite network. The chapters will also cover the effects of atmospheric turbulence and scattering effects to develop high-speed optical routing techniques via appropriate route for both satellite and terrestrial paths to optimize the communication channel links to provide the high availability and lowest latency under any weather and atmospheric conditions at all times. The rules of the game for a *single satellite* and a *constellation of satellites* are different because the laser communication link issues, problems, and requirements are different. Other topics are discussed in the next few chapters: home/indoor communications, outdoor including terrestrial and space-based terminals, satellite-to-cell phone, all-optical technology including advanced developments with Li-Fi and visible light communication (VLC), Li-Fi hotspots, advanced optical networks to connect all the terminals/platforms, last-mile problem solution with free-space laser communications, and a combination with high-speed optical fibers (but still using optical wavelength). Finally, the most recent developments of compact (shoebox size) terahertz laser to complement small cells and microcells approach for fixed and moving terminal situations will also be discussed. Future applications like the following will also be mentioned: (1) wireless world in 2050 and beyond: a window in the future; (2) what is beyond 5G: future look at 6G and beyond for global and rural connectivity connecting the remaining 4 billion; (3) solutions for future smart city concepts; (4) global Internet of Things (IoT) and machine-to machine (M2M): worldwide possibility with optical technology; and (5) advanced optical networks in designing these to make this connectivity for the end users.

This chapter therefore links to next chapters with the goal of establishing a global laser satellite constellation.

1.5 Summary

This chapter discusses the basics of satellite wireless communications, which is an essential link for a connected world. Fundamental concepts of satellite communica tions with emphasis on basic metrics for satellite communications, and link performance for communications between a satellite and a ground station as well as for inter-satellite links (ISL), are explained. Lead-ahead angles for establishing a link between a ground station and a satellite, between two LEO satellites and between two GEO satellites, are presented. Doppler shifts for satellite communication systems and potential compensation techniques relevant to GEO, LEO, and lower orbit satellites are addressed, which are essential in designing and developing satellite communication systems. Various satellite orbits and different types of satellites at different orbits are compared. Single-satellite *all-optical* concept technology for providing global broadband connectivity is introduced. Finally, a communication architecture (three-layer networks) to integrate satellite communications into terrestrial and other satellite networks using laser/optical (FSO) links is discussed, and the concept of satellite constellation networks is introduced.

References

1. Arun K. Majumdar, *Optical Wireless Communications for Broadband Global Internet Connectivity: Fundamentals and Potential Applications*, Amsterdam, Netherland Elsevier 2019
2. Ghassemlooy Z. and Papoola, W.O., Terrestrial Free-Space Optical Communications, Mobile and Wireless Communications Network Layer and Circuit Level Design, Salma Ait Fares and Fumiyuki Adachi (Ed.), ISBN: 978-953-307-042-1, InTech, 2010. https://en.wikipedia.org/wiki/Space_law

3. "Space Law", *United Nations Office for Outer Space Affairs.* http://www.unoosa.org/oosa/en/ourwork/spacelaw/index.html
4. Arun K. Majumdar, Chapter 1 and Chapter 2, *Advanced Free Space Optics (FSO): A Systems Approach*, Springer Science+Business Media, New York 2015.
5. Gerard Maral and Michel Bousquet, *Satellite Communications Systems: Systems, Techniques and Technologies,* 5th Edition, John Wiley, 2009.
6. Low-Earth-Orbit Satellites and IoT: The Next Space Revolution, Avnet Silica, Aerospace & defense, June 27, 2019.
7. Catalog of Earth Satellite Orbits, NASA earth observatory, a part of EOS Project Science Office at NASA Goddard Space Flight Center: https://earthobservatory.nasa.gov/features/OrbitsCatalog
8. Do Xuan Phong, A Feasibility Study for Laser Communications between Micro Satellites and GEO Satellites, Master's Dissertation, Keio University, September 2015.
9. Frigyes, Istvan, Janos Bito, and Peter Bakki, *Advances in mobile and wireless communications:* views of the 16th IST mobile and wireless communication summit, Vol. 16, Springer Science & Business Media, 2008.
10. Inigo del Portillo, Bruce G. Cameron, Edward F. Crawley (Massachusetts Institute of Technology), *A Technical Comparison of Three Low Earth Orbit Satellite Constellation Systems to Provide Global Broadband,* October 1st 2018, 69th International Astronautical Congress 2018 Bremen, Germany, IAC-18-B2.1.7.
11. Richard Welle, Alexander Utter, Todd Rose, Jerry Fuller, Kristin Gates, Benjamin Oaks, and Siegfried Janson, *A CubeSat-Based Optical Communication Network for Low Earth Orbit*, 31st Annual AIAA/USA Conference on Small Satellites SSC17-XI-01
12. International Regulations for Nano/Pico Satellites, ITU Seminar, 14-16 April 2014, Cyprus https://www.itu.int/en/ITU-R/space/workshops/cyprus-2014/Documents/Presentations/Tony%20Azzarelli%20-%20Nanopicosatellites.pptx
13. https://www.comsoc.org/education-training/training-courses/online-courses/2020-09-optical-wireless-communications-recent
14. Arun K. Majumdar and Jennifer C. Ricklin, "Free-Space Laser Communications: Principles and Advances," Springer, New York 2008.
15. Daniel Nieto Yll, "Doppler shift compensation strategies for LEO satellite communication systems," A degree thesis, Universitat Politècnica de Catalunya, Barcelona, June 2018.
16. Ifran Ali, Naofal Al-Dhahir, John E. Hershey. "Doppler Characterization for LEO Satellites" IEEE Transactions on communications, vol 46 no. 3, pp. 309-313, March 1998.

Chapter 2
Free-Space Optical Propagation Relevant to Integrated Space/Aerial, Terrestrial, and Underwater Links

2.1 Introduction

For free-space optical communication (FSOC) link between an optical ground station (OGS) and a satellite such as geostationary (GEO), low Earth (LEO), small, or even microsatellite in lower altitudes involves a slant path connecting the ground, satellite, or airborne platform. From an optical propagation point of view, the uplink and downlink paths are very different because the turbulence and scattering layer is mostly near the transmitter on an uplink path and only near the receiver on a downlink propagation path. Thus, main problems appear in the uplink where the optical beam is spread much because of the nature of spherical wave propagation. The downlink wave can be treated in general as approximately plain-wave propagation. Accurate and reliable modeling for both effects simultaneously is needed for designing high-data-rate robust optical communication systems. There are excellent references available for optical propagation through atmospheric turbulence and scattering medium. Some of the recent references can be found in Refs. [1–4]. Optical propagation effects most relevant to optical communication are scintillation effects like *beam spreading, scintillation index, spatial coherence radius of the wave, coherence time*, and their effects on *probability of fades*. However, there are some fundamental differences and more complex issues in developing accurate atmospheric channel model for satellite or other airborne platform for maintaining reliable optical communication. Some of these are below:

1. Satellite or other airborne platforms are moving terminals. The propagation slant-path modeling needs to take into account the variations of propagation parameters, if any, for moving terminal.
2. For constellation of satellites, the predicted atmospheric channel modeling should be adjusted during the so-called handoff/handover and seamless operation to represent the moving dynamic slant paths.
3. Anisoplanatic nonreciprocity effects between uplink and downlink.

© Springer Nature Switzerland AG 2022
A. K. Majumdar, *Laser Communication with Constellation Satellites, UAVs, HAPs and Balloons*, https://doi.org/10.1007/978-3-031-03972-0_2

4. Anisotropic atmospheric turbulence statistics along nonhomogeneous path, specially turbulence in the upper troposphere and stratosphere, may follow *non-Kolmogorov* power spectrum.
5. Atmospheric channel model can be different for ground-to-satellite and inter-satellite optical paths. This in turn makes a difference in the design of FSO communication systems for reliable pointing, acquisition, and tracking (PAT).
6. For exploring under water communications involving satellite and underwater optical terminals, a combined model to include atmospheric channel and random air-water interface is necessary.
7. Fog near the ground terminals can severely limit the FSO communications capability and the maximum data rate transmission.
8. Switching to a different OGS to avoid interruption by clouds requires the FSO communication design to deal with two different atmospheric communication channels.

In this chapter, the current existing atmospheric models are examined as applied to ground-satellite uplink/downlink slant paths. In particular, the atmospheric effects relevant to designing an FSO communication system performance are discussed. The fundamental difference between the horizontal model and the slant-path model is that in the latter one needs to consider the changes of refractive index structure parameter C_n^2 along the path. Typically, an existing profile model of C_n^2 as a function of altitude is modified including a zenith angle parameter. However, the conventional models are valid for a zenith angle less than 60° (valid for Rytov approximation) or less than 45° for large ground-level C_n^2. For designing a constellation of satellites for uninterrupted global optical communication, it is also necessary to design the minimum elevation angle which is defined as the minimum allowable angle between the horizon and any viewable spacecraft (e.g., satellite) at a ground point. The purpose of introducing this angle is to account for terrain obstructions, such as buildings (including very tall skyscrapers), mountains, and trees. Some typical values for minimum elevation angles may be 5–10° for 24-h constellation [5]. The validity of the conventional profile model for small elevation angles is questionable. It may therefore be necessary to use strong fluctuation models to account for beam wander effects with large-diameter beams to apply for large zenith angles similar to the near-horizon propagation paths [4].

This chapter discusses the atmospheric models relevant to characterizing and analyzing the free-space optical communication technology for designing reliable communication systems which operate between ground and spacecraft (including satellite) as well as for inter-satellite links. Examples of different link scenarios for achieving all-optical technology for global connectivity are discussed.

2.2   Review and Summary of Atmospheric Propagation Parameters' Results

This section reviews, discusses, and summarizes the essential results of some atmospheric effects that are of importance in the design of any satellite FSO communication system. Results specifically applicable to other spaceborne terminals such as HAP and UAV operating at lower altitudes are also discussed with necessary working equations needed for designing a communication system. The difference in uplink and downlink paths reflects the derived statistical quantities and expressions for beam wandering (spreading), scintillation index, and spatial coherence radius, all of which have an effect on the probability of fade. A summary of the mathematical expressions, already published and are available in literature, will be listed in a later section. For slant-path communication links, the beam wander effects relevant to "fading times" are more in uplink and not in downlink.

Note that the main challenge for FSO communication is the propagation channel (distance between FSO terminals), which can include *atmosphere* (turbulence, clouds, diffusion, and transmission) and *aerodynamic perturbations* (when used aircraft as a terminal). These are complex contributors but are the key for system availability and have major effects on several key optical communication performances such as optical quality, pointing, and transmission. A reliable propagation channel modeling is therefore the key to guarantee link performances through the atmosphere. Laser communication through dense clouds is not practically possible because of extremely high propagation loss. One possible solution to overcome clouds issue might be to develop *site diversity* with a network of *de-correlated* ground stations to achieve FSO communication system availability, which obviously requires to select optimized location of ground station. Figure 2.1 shows an illustration of atmospheric influence on laser communication between OGS and satellite, HAP, and UAV space-based optical communication terminals for uplink and downlink slant paths.

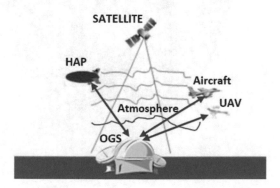

Fig. 2.1 An illustration of atmospheric influence on laser communication between optical ground station (OGS) and satellite, HAP, and UAV space-based optical communication terminals for uplink and downlink various slant paths

To mitigate the effects of atmospheric turbulence due to uplink or downlink slant path, links for developing FSO communication systems, sometimes adaptive optics (AO) techniques, are used where the phase distortions of a known source (reference) are sensed first and then a corrective or conjugate phase to the outgoing beam is applied. An *isoplanatic angle* is defined as the angular distance from the reference (which, for example, can be a beacon) over which atmospheric turbulence characteristics remain essentially the same. Another related propagation effect is the *angular anisoplanatism*, which occurs when a propagating beam is offset by a constant angle from the direction along which the turbulence is measured. The atmospheric turbulence [4]. The angular anisoplanatism is important for a moving uncooperative satellite target when two beams are used to track, one to track a moving uncooperative satellite target and the second beam used to intercept the target in a point-ahead configuration. The concept of a lead-ahead (or point-ahead) angle for a ground station and a satellite was discussed in the previous chapter. Figure 2.2 shows a geometry where both the point-ahead angle and the isoplanatic angles affecting uplink OGS to a satellite for FSO communication are shown. The point ahead is much larger than the isoplanatic angle as shown in the figure.

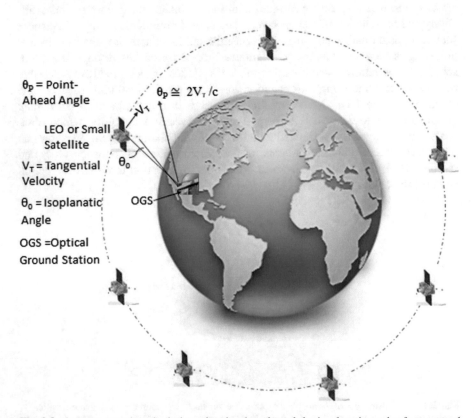

Fig. 2.2 A geometry where both the point-ahead angle and the isoplanatic angles for a ground station to a moving satellite are shown for comparison

Note that both the ground and space terminals must be equipped with laser/optical transceivers for accomplishing two-way (duplex) communications. The goal will be to establish high-data-rate communications between any location on the Earth and any other locations including remote areas *anywhere anytime*. Ultimately, this technological concept of *all-optical* technology will be having the potential to provide broadband global Internet connectivity.

One example of considering a point-ahead angle for satellite-based quantum communication is the fact that the point-ahead angle caused by satellite motion is several times larger than the required extremely small divergence angle of the signal beam, and therefore the transmitting quantum entangled photon beam must compensate for the point-ahead effect due to relative motion of the moving satellites. Figure 2.3 depicts a sketch to show the absolute necessity for compensating the point-ahead angle due to satellite motion applicable to satellite quantum communication uplink configuration. This is absolutely essential in designing satellite-based quantum communication system. This is also true for developing any FSO space communication system between moving platforms. From an optical propagation point of view, however, the line-of-sight (LOS) dynamic optical channel model also needs to be considered because the turbulence distributions are different at the beginning and at the end of the point-ahead angle for the transmitter to reach to the receiving antenna. The transmitter therefore needs to offset with a small point-ahead angle to pre-point the next location.

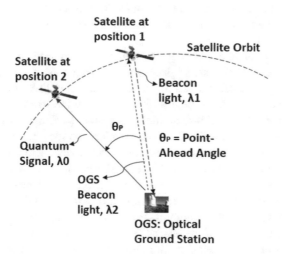

Fig. 2.3 A sketch to show an example of the point-ahead angle due to satellite motion applicable to satellite quantum communication uplink configuration

2.2.1   Atmospheric Channel and Its Role in FSO Communication

To design a reliable FSO communication system to operate successfully in the presence of atmospheric turbulence and scattering medium, it is necessary to understand the optical propagation effects through these various slant paths on laser/optical communication. Optical propagation through random medium, in particular turbulence, has been a subject of research for the last number of decades. Tatarskii's original work [6] translated from Russian has been a pioneering research work applicable to fundamental concepts of the theory of turbulence.

Atmospheric turbulence limits the reliable optical communication channels in establishing high-capacity, high-data-rate optical communication for both horizontal paths and for those designed to link with GEO, LEO, or small satellite. *Scintillation* is caused by random variations of atmospheric refractive index due to microscale or small eddies in the atmosphere. These atmospheric effects are important in designing a reliable and efficient FSO communication system and evaluating the system performance. There are a number of excellent research works published in the last few years and are available in the literature [1–4, 7–12]. The results and the detailed derivations of atmospheric parameters relevant to optical communications are available and will not be repeated in this chapter. The interested readers will find the references useful in developing a background knowledge about understanding the interactions of atmospheric turbulence and scattering medium with optical wave [7–12].

When larger turbulence eddies in the atmosphere crossed by the entire optical beam cause the beam to deviate slightly from its original path, it causes *beam wander*. The beam wander affects "fading times" more in uplink and not as much in downlink. The beam wander is negligible in the downlink from satellites to ground station for evaluating optical communication system performance. For uplink, the beam width during its atmospheric path is smaller than the outer scale of the turbulence and changes the outgoing direction significantly when leaving the atmosphere. For downlink, turbulence much smaller than the beam size does not displace the beams' centroid significantly. When downlink reaches the atmosphere close to the atmosphere, the effect of waveform tilt primarily causes angle-of-arrival fluctuations. For a two-way optical communication where both uplink and downlink channels are involved, it is therefore necessary to understand the combined effects of both scintillation and beam wander. The results of these effects determine the design of FSO communication system in order to operate under atmospheric turbulence and scattering media.

2.2.2   Various FSO Transmitting Beam Shapes and Geometries Relevant to Long-Distance Data Transmission

For FSO communication systems between a ground terminal and spaceborne/airborne terminals, the message at first has to be modulated by a carrier generated by the transmitter which is a laser or other optical source. Typically, plane wave, spherical wave, or Gaussian beam wave (fundamental mode) excitations at the source, i.e., at the transmitter, have been extensively studied. The atmospheric turbulence effects such as scintillations are then calculated to determine the communication system parameters like signal-to-noise ratio (SNR), bit error rate (BER), and others. Because of the recent tremendous growth in the telecommunications infrastructure, using of FSO communication links requires different types of optical networks and scenarios for various applications such as global Internet connectivity, Internet of Things (IoT), and 5G. For these optical links to be efficient and reliable for searching for the best and optimum beam at the source (transmitter), it is important *to use a beam to minimize the degrading effects induced by the atmospheric turbulence.* Figure 2.4 illustrates some of the typical transmitting beam types and their shapes, which are commonly used for FSO communications for both horizontal and slant-path links.

2.2.3   Review and Summary of Intensity Profiles of Various FSO Transmitting Beams

Mathematical formulations describing the intensity profiles of FSO transmitting beams propagated through atmospheric turbulence are already discussed elsewhere [2]. Because of the recent demands of tremendous amount of data requirement using cell phones and laptops to be connected with Internet from almost anywhere, anytime, the optical networks and scenarios may have different types of FSO transmitting beam shapes other than the conventional plane, spherical, or Gaussian beam waves. In order to search for the best optimum beam at the transmitting source, the main criteria will be *to use a beam that minimizes the degrading effects induced by atmospheric turbulence.* A brief summary of the intensity profiles is provided below. In order to analyze the FSO communication system performance, the received intensity profiles at the receiver end are required to evaluate parameters such as SNR, BER, and probability of fades. When the optical wave propagates through a random medium, it undergoes both amplitude and phase fluctuations when the signal arrives at the detector surface.

2.2.3.1   Plane Wave

The complex amplitude at a communication distance z from the transmitter wave propagating along the positive z-axis in free space:

$$U_0(r,z) = A_0 e^{i\phi_0 + ikz} \qquad (2.1)$$

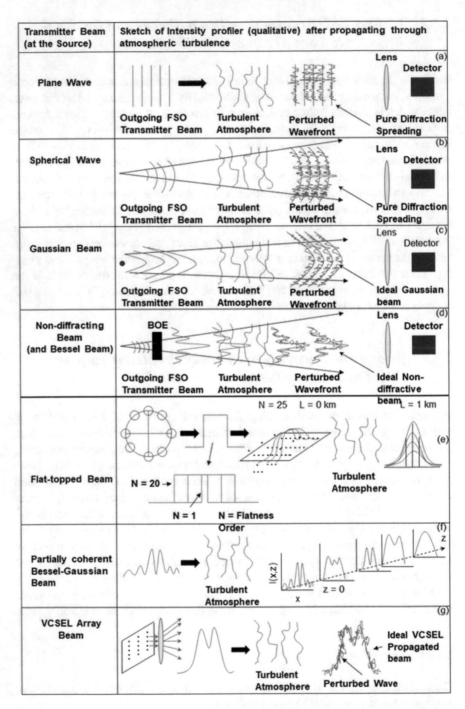

Fig. 2.4 Sketches to illustrate intensity profiles of transmitting beams propagated through atmospheric turbulence. (*Reprinted with permission from Springer Nature and Copyright Clearance Center, 2021* [2])

where A_0 is a constant amplitude of the field, ϕ_0 is the phase, and k is the optical wave number and is equal to $2\pi/\lambda$.

2.2.3.2   Spherical Wave

For FSO communication system, the source excitation of a spherical wave can be considered as a point source. The complex wave amplitude at a distance z is given by

$$U_0(r,z) = \frac{A_0}{4\pi z} \exp\left(ikz + \frac{ikr^2}{2z} \right) \tag{2.2}$$

The phase with a transverse radial dependency is

$$\phi = k\left(z + \frac{r^2}{2z} \right)$$

2.2.3.3   Gaussian Beam Wave

Most commonly used, the complex amplitude at a distance z from the source a Gaussian wave [4] at a radial distance r from the beam center line is given by

$$U_0(r,z) = \frac{1}{1+i\alpha_0 z} \exp\left[ikz - \frac{1}{2}\left(\frac{\alpha_0 k}{1+i\alpha_0 z} \right) r^2 \right] \tag{2.3}$$

The complex parameter α_0 is related to spot size and phase front radius of curvature by the following equation [4]:

$$\alpha_0 = \frac{2}{kW_0^2} + i\frac{1}{F_0}$$

where W_0 is the radius at which the field amplitude is at $1/e$ of that on the beam axis, and F_0 is the radius of curvature of the phase front (parabolic). For collimated beam $F_0 = \alpha_0$, for convergent beam $F_0 > 0$, and for divergent beam $F_0 < 0$.

2.2.3.4   Non-diffractive Beam

Non-diffractive beam is an ideal optical field with completely eliminated diffraction and has been of current research in applying to the long-range FSO communication that has to operate under atmospheric turbulence. A realizable approximation of ideal non-diffractive beam and pseudo-non-diffractive (P-N) beam and a number of ways to generate non-diffractive beams are discussed [21]. A *binary optical element* (BOE) to shape incident plane wave or divergent spherical wave into a non-diffractive beam is also discussed [20]. In a typical long-distance FSO communication system design, the choice of selecting non-diffractive wave will also depend on the complexity and the costs involved.

The intensity distribution of a non-diffractive beam for a spherical wave at a distance z from the BOE element, and for the BOE located at the $(x, y, 0)$ coordinate $(z = 0)$, is given by [20]

$$I(\sigma,\phi,z) = \frac{A^2}{\lambda^2 z^2}$$

$$\left| \int\int_{-\infty}^{\infty} \exp\left(j\frac{k\rho^2}{2f} \right) x \exp\left(j\frac{\pi R^2}{\lambda(d_2-d_1)} \right) \ln\left(\left(\frac{z_{d_2}-z_{d_1}}{R^2} \right) d_1 + \left(\frac{d_2-d_1}{R^2} \right)^2 \rho^2 \right) \right|^2$$

$$\left| \exp\left(j\frac{k}{2z}\left[\rho^2+\sigma^2-2\rho\sigma\cos(\theta+\varphi) \right] \right) \rho d\rho d\theta \right| \qquad (2.4)$$

where A is the constant amplitude of the initial wave amplitude, the wave number $k = 2\pi/\lambda$, the optical wavelength is λ, f is the distance between the source point and BOE element, R is the radius of the element, and d_1 and d_2 are points on the optical axis, non-diffractive beam, propagating between z_{d_1} and z_{d_2}.

2.2.4 Recently Studied Other Types of Beams for Long-Distance Applications

For long-distance FSO communication applications such as ground-to-space terminals (satellite, aerial, HAP, UAV), there are a number of researches recently published describing various types of transmitting beam shapes to optimize the signal power received at the receiving terminal when propagating through atmospheric turbulence. Some of the beam shapes are already sketched in Fig. 2.4. This section discusses some of the different types of beam shapes, which can be useful for optimizing FSOC system design to operate in the presence of atmospheric turbulence. One of the major criteria to choose an optimum beam is to minimize the beam wandering effects so that angle-of-arrival fluctuations are small to help in designing pointing and tracking system.

Researchers recently studied other types of transmitting beams [13, 19], which are discussed briefly below. These different types of beams have various propagation characteristics when transmitted through atmospheric turbulence. For efficient design of a FSO communication system which has to operate in the presence of atmospheric turbulence, the optimum choice of the type of transmitting beam will finally depend on the propagation characteristic of the outgoing beam, the received power requirement at the receiver, and the complexity of the transmit and receive optics.

2.2.4.1   Bessel Beam

A non-diffracting, self-healing Bessel beam [13] can potentially mitigate the problems of atmospheric turbulence and diffraction for FSO data transmission. The intensity profile of Bessel beams is cylindrically symmetrical with a central core surrounded by a set of concentric rings.

Properties of Bessel beams are very promising for FSO communication between OGS and space-based terminals. A Bessel beam of the 0th order can be represented by [15]

$$E(r,\phi,z) = A_0 \exp(ik_z z) \int_0^{2\pi} \exp\left[ik_r(x\cos\phi + y\sin\phi)\right]\frac{d\phi}{2\pi} \qquad (2.5)$$

where k_r is the radial wave vector and k_z is the longitudinal wave vector with the radial coordinate r and $r^2 = x^2 + y^2$, and $k = \left(k_z^2 + k_r^2\right)^{1/2} = 2\pi/\lambda$. The angle of the cone defined by the Bessel beam's wavefront is

$$\theta = \arctan\left(\frac{k_r}{k_z}\right) \qquad (2.6)$$

The beam radius is then $b_w = 2.405/k_r$. This Bessel beam can propagate to a maximum distance, given by

$$z_{max} = \frac{B_w k}{k_r} = \frac{B_w}{\tan\theta} \qquad (2.7)$$

The power in the Bessel beam up to a radius b is [15]

$$P = A_0^2 \int_0^{2\pi}\int_0^b J_0^2(rk_r)\,r\,dr\,d\phi = A_0^2 b^2 \pi \left(J_0(k_r b)^2 + J_1(k_r b)^2\right) \qquad (2.8)$$

The power from the central peak up to the first zero is given by

$$P_{B_w} = A_0^2 B_w^2 \pi J_1(2.4048) = 0.8467 A_0^2 B_w^2 \qquad (2.9)$$

The above parameters are useful in designing FSO communication system using a Bessel transmitting beam. The above equations for the power at a receiver distance will determine the received SNR needed for calculating the BER and probability of fades. If the Bessel beam can be produced efficiently, for long-distance propagation for FSO communications Bessel beams have a number of advantages over Gaussian beam profiles in terms of power delivery [14, 15, 17].

2.2.4.2   Aperture-Truncated Airy (Anti-Airy) Beam

One of the novel beams to improve the receiving performance, recently the aperture-truncated (anti-Airy) beam, is proposed which is based on the fundamental properties of Airy beam with features of non-diffracting propagation in free space, self-accelerating in the transverse direction and self-healing in the presence of obstacles. These features are promising for applications in FSO communication operating in the presence of atmospheric turbulence since the power loss due to diffraction spreading can be minimized in long-haul links such as ground-to-satellite. In practical FSO communication system, the transmitting beam is always truncated by the finite transmitter aperture, and for a limited distance the Airy beam can propagate within a limited nearly diffraction-free range [16]. For optical carrier used in FSO communication, the well-preserved self-accelerating feature of finding a way around obstacles such as clouds or other objects makes it more robust for a communication link such as satellite-to-ground communication using Airy beams than the existing line-of-sight (LOS) link. For FSOC, the transmitter (TX) optics should be larger than the spatial extent of the Airy field, determined by $k^2 x_0^2$ where k is the optical wave number and x_0 is the characteristic length of the Airy function. Transmitter aperture size will require to properly choose the value of the characteristic length of the Airy function, x_0, for designing a communication system for effective capture and tracking. A 2D ideal Airy beam can be written as follows [16]:

$$U\left(s_x, s_y\right) = Ai\left(s_x\right)\exp\left[i\upsilon s_x\right]Ai\left(s_y\right)\exp\left[i\upsilon s_y\right] \tag{2.10}$$

where the characteristic lengths normalized in both directions x and y are given by

$$s_x = x \,/\, x_0$$

and

$$s_y = y \,/\, x_0$$

For FSOC systems, there is a finite TX aperture size which is also a natural boundary. Aperture-truncated ideal Airy (anti-Airy) beam is an ideal Airy beam truncated by the TX optics. For an anti-Airy beam, the field aperture with R as the radius of the TX aperture can be expressed as

$$U_{\text{anti}}\left(s_x, s_y\right) = Ai\left(s_x\right)\exp\left(i\upsilon s_x\right)Ai\left(s_y\right)\exp\left(i\upsilon s_y\right) \times \text{circ}\left(R\right). \tag{2.11}$$

Figure 2.5 shows the intensity distributions of an ideal, untruncated Airy beam and an anti-Airy beam.

The anti-Airy beam, i.e., an ideal Airy beam with no additional truncation, is promising for long-distance FSOC links.

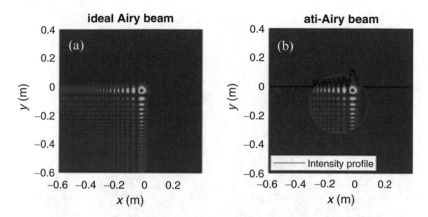

Fig. 2.5 Intensity distributions of an ideal Airy beam (**a**) without truncation and an aperture-truncated Airy beam (**b**). (*Reprinted with permission from publisher IARIA SPACOMM 2018* [16])

2.2.4.3   Hermite-Gaussian Beam

Recently, Hermite-Gaussian beam for the application of optical wireless communication has been investigated [22], which can be helpful in improving the system performance of FSOC communication system from Earth to satellites or from satellites to Earth terminals. The field of the higher order of Hermite-Gaussian modes of a laser can be written as [22]

$$u_s(x,y) = H_n(x/\alpha_s)H_m(y/\alpha_s)\exp\left[-\left(0.5x^2/\alpha_s^2\right)\right]\exp\left[-\left(0.5y^2/\alpha_s^2\right)\right] \quad (2.12)$$

where (x, y) is the transverse coordinate at the source plane, α_s is the Gaussian source size, $H_n(\cdot)$ and $H_m(\cdot)$ are the Hermite polynomials in the x and y directions for the mode numbers, n and m are in the x and y directions. The Hermite-Gaussian beams suffer less than the beam spreading caused by Gaussian beams [18]. Figure 2.6 shows the longitudinal distribution of the Hermite-Gaussian beam for the mode numbers m, $n = 2$, and clearly indicates that the shape of the beam is not changing when propagating in vacuum (without turbulence) [18].

2.2.4.4   Bessel-Gaussian Beams (BRNO UNIV)

Bessel-Gaussian beams can find applications for ground-to-satellite or satellite-to-ground FSO communication because of their non-diffractive properties as the optical beam propagates through atmospheric turbulence. Field of the Bessel-Gaussian beam is given by [18]

$$E(s, z = 0, \omega) = J_n(\alpha s)\exp(in\theta)\exp\left(-\frac{s^2}{w_0^2}\right) \quad (2.13)$$

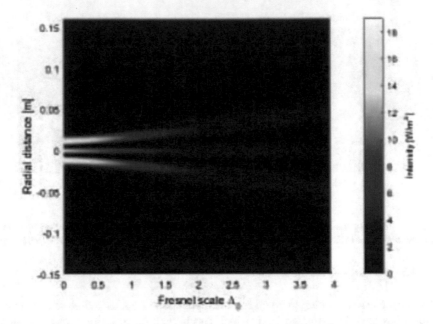

Fig. 2.6 The longitudinal distribution of Hermite-Gaussian beam (mode numbers $n, m = 2$) propagating through vacuum (no turbulence). (*Courtesy of and Copyright permission to reproduce from Prof. Peter Barcik, BRNO University, 2021* [18])

where n denotes the topological charge, J_n is the Bessel function, α and w_0 are the width parameters, w is the angular frequency, and s is the transverse vector. Figure 2.7 shows the field distributions for the first kind of Bessel-Gaussian beams and a comparison with propagation through vacuum (no turbulence). Although for weakly turbulent atmosphere scintillation index is lower than the scintillation index of the Gaussian beam, for long-distance propagation and large beam width, the scintillation index is higher than the Gaussian beam.

2.2.4.5 Laguerre-Gaussian Beams (BRNO UNIV)

A Laguerre-Gaussian beam at the transmitter can be described as follows [18, 23, 24]:

$$U(s,\phi,z=0) = \left(\frac{\sqrt{2}s}{w_0}\right)^l L_p^l\left(\frac{2s^2}{w_0^2}\right)\exp\left[-\frac{s^2}{w_0^2} + il\phi\right] \qquad (2.14)$$

where L_p^l is the associate Laguerre polynomial, w_0 is the radius of the beam, and (s, ϕ) is the cylindrical coordinate system. Scintillation index of this beam is lower than the scintillation index of the Gaussian beam. In designing the FSOC system, if

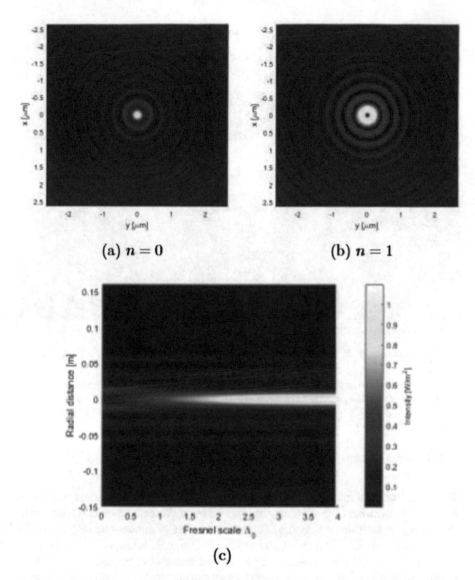

Fig. 2.7 Field distributions for the first kind of Bessel-Gaussian beams: (**a**) $n = 0$, (**b**) $n = 1$, (**c**) longitudinal intensity distribution for $n = 1$ for propagation through vacuum (no turbulence). (*Courtesy of and Copyright permission to reproduce from Prof. Peter Barcik, BRNO University, 2021* [18])

the parameters $l = 0$ and p as high as possible are chosen, then the lowest scintillation can be archived. Figure 2.8 [18] shows the results from simulated field distribution of the Laguerre-Gaussian beam for different values of the parameters l and p defined in Eq. (2.14).

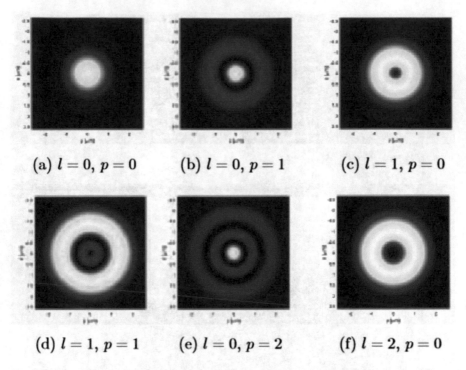

(a) $l = 0, p = 0$ (b) $l = 0, p = 1$ (c) $l = 1, p = 0$

(d) $l = 1, p = 1$ (e) $l = 0, p = 2$ (f) $l = 2, p = 0$

Fig. 2.8 Results from a simulated field distributions of the Laguerre-Gaussian beam for different values of the parameters, l and p: (**a**) $l = 0, p = 0$, (**b**) $l = 0, p = 1$, (**c**) $l = 1, p = 0$, (**d**) $l = 1, p = 1$, (**e**) $l = 0, p = 2$, (**f**) $l = 2, p = 0$. (*Courtesy of and Copyright permission to reproduce from Prof. Peter Barcik, BRNO University, 2021* [18])

2.2.5   Atmospheric Channel Models for FSO Communication

There has been a tremendous research in understanding the atmospheric effects of FSO communication for both terrestrial and between ground station and spaceborne terminals like satellite or airborne. There are a variety of deleterious features of the atmospheric channel that may lead to serious FSO signal fading, and even the complete loss of signal altogether. Since FSO communication requires LOS optical beam transmission and reception, other factors can cause physical obstructions such as birds and tree limbs to temporarily or even permanently block the FSO LOS signals. Other effects like physical/building motion due to wind or ground motion can sometimes cause misalignment of fixed-position FSO communication systems. The two main factors which cause the FSO transmitted optical to decrease are absorption and scattering due to particulate matter such as aerosols, dust, smoke, haze, and fog, and intensity fading and random signal losses at the receiver due to optical turbulence.

For FSO communication applications, an optical beam propagates through the atmosphere above the ground and basically a mixture of gases, molecules, and particles that continuously gain or lose energy (heat). Because of the heating of the

ground by the Sun and wind speeds, there is a constant movement of air cells causing thermal turbulence in air cells characterized by inhomogeneous and dynamically changing refractive index at microscales, density, and air consistency. During
daytime, the *atmospheric boundary layer*, which generally extends to roughly
1–2 km above the Earth's surface, and atmospheric dynamics are dominated by the
interaction and heat exchange with the Earth's surface. This boundary layer is also
difficult to model exactly for accurate evaluation of FSO signal transmission. The
optical turbulence is strongest near the ground which can be due to the fact that the
convection dominates the boundary layer dynamics, and the convective instability
gives rise to thermal plumes and strong turbulence. Around sunrise and sunset, the
surface and air temperatures are very nearly identical resulting in very low wind
speed which causes small values of turbulence. During night, warmer air over colder
ground creates more stable conditions. FSO light beams are severely affected by the
atmosphere, which includes change in polarization, refraction, absorption, scattering, and attenuation. This results in random fluctuations of the transmitting and
receiving light beam at a frequency within a range of about 10 mHz and 200 Hz [25]
or more. When the laser beam interacts with atmospheric turbulence, its polarization and coherency fluctuate due to random fluctuations of the air mas along the
LOS, resulting in constantly fluctuating attenuations resulting in non-consistent
power loss throughout the air mass along the path. In an optical wireless communication system, the intensity of the received signal fluctuates due to random temporal
and spatial irradiance fluctuations of the optical beam. The signal therefore is
focused and defocused onto the photodetector randomly. The random signal fluctuations due to atmospheric turbulence are generally termed as *scintillation* and are
measured as the strength of turbulence which is normally denoted by C_n^2. For high
data rate like multi-Gbit/s long-distance FSOC communication (ground-to-satellite
or airborne) systems to operate in the presence of atmospheric turbulence, scintillations of a laser/optical beam are a major problem which ultimately limit the efficient
data transmission and reception. While absorption and scattering due to atmospheric
particulates may significantly decrease (attenuate) the transmitted signal, optical
turbulence can severely degrade the wave-front quality of signal-carrying laser
beam, causing intensity fading and random signal losses at the receiver. It is therefore necessary to take into account simultaneous effects of both the signal attenuation and random fluctuations in order to analyze the FSO communication system
performance.

2.2.5.1   All-Optical Communication Channels for Various Optical
Communication Architecture and Links

This section discusses some configurations of space/aerial/terrestrial/underwater
free-space optical (FSO) communication links that show how atmospheric turbulence and scattering media impact laser communication. The optical channel models are different for the following different terminals:

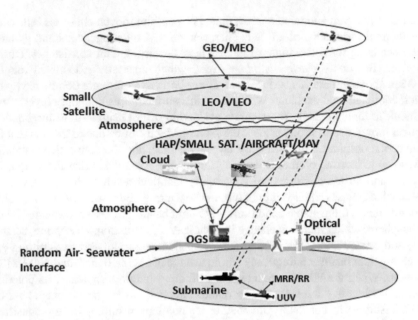

Fig. 2.9 Sketch of integrated system architecture for space/aerial/terrestrial/underwater terminals showing site diversity for laser communication links

1. Satellite-to-ground, HAP, aircraft or UAV-to-ground
2. Inter-satellite links (ISLs)
3. Very-low-altitude small satellite-to-ground showing 5G/6G potential
4. Shore-to-underwater and underwater links (using also modulated retroreflector (MRR))

Figure 2.9 shows integrated system architecture for space/aerial/terrestrial/underwater terminals showing site diversity for laser communication links propagated through space, atmosphere, air-water interface, and water. Each communication link is different by itself in terms of optical propagation characteristics affecting FSO communication system performance.

2.2.5.2 Optical Turbulence Model Descriptions Relevant to FSO Communication

For ground-to-satellite communication propagation along a vertical or slant path, a $C_n^2(h)$ profile from the height of the receiver above sea level to the top of the atmosphere for up to about 40 km is required. However, the $C_n^2(h)$ profile changes with time during ground-to-satellite FSO communications. Some excellent empirical models, including both day- and nighttime models applicable to ground-to-space or space-to-ground FSO communication, are available in the literature. The models are based on experimental measurements at various locations, time of day, and wind speeds.

Optical turbulence models	Model description and the physics involved	Applicability for FSO communication
PAMELA model [11, 26, 27]	Provides estimates of $C_n^2(h)$ within the surface boundary layer and the required inputs are latitude, longitude, date, time of day, percent cloud cover, and terrain type. For slant-path measurements at the height of the laser, receiver should be used for accurate estimate of $C_n^2(h)$. The calculations involve solar insolation, irradiance, heat flux, Pasquill stability, Monin-Obukhov length, and flux profile relationships. The details of the derivation of PAMELA model are discussed in the references. The value of turbulence strength is calculated from $$C_n^2 = \frac{bK_h}{\varepsilon^{1/3}}\left(\frac{\partial n}{\partial h}\right)^2,$$ where b is a constant commonly approximated by 2.8, K_h is the turbulent exchange coefficient for heat, ε is the energy dissipation rate.	– Slant path including different terrains and weather conditions – Sensitive to wind speed and the necessary meteorological inputs
Hufnagel-Valley (H-V) model [4, 28, 29]	$$C_n^2(h) = 0.00594\left[\left(\frac{V}{27}\right)^2 (10^{-5}h)^{10}\exp\left(\frac{-h}{1000}\right)\right.$$ $$+2.7\times10^{-16}\exp\left(-\frac{h}{1500}\right)$$ $$\left.+A\exp\left(-\frac{h}{100}\right)\right]m^{-2/3},$$ where h is the altitude (m), V is the rms windspeed (m/s), and A is the nominal value of $C_n^2(0)$ at the ground (m$^{-2/3}$). $$A = 1.29\times10^{-12}r_0^{-5/3}\lambda^2 - 1.61\times10^{-13}\theta_0^{-5/3}\lambda^2 - 3.89\times10^{-15}$$ θ_0 is the isoplanatic angle and r_0 is the coherence length. $$V = \left[\frac{1}{15}\int_{5\times10^3}^{20\times10^3}v^2(h)dh\right]^{1/2}$$ where $v(h)$ is the Bufton wind model given by $v(h) = w_s h + v_g + 30\exp[-\{(h-9400)/4800\}^2]$ where v_g is the ground wind speed and w_s is the slew rate associated with a satellite moving with respect to an observer on the ground.	

Optical turbulence models	Model description and the physics involved	Applicability for FSO communication
HAP model (see Larry's work at China Lake) [4, 30, 31]	HAP model is a variation of the HV model: $$C_n^2(h) = M\left[0.00594\left(\frac{v}{27}\right)^2\left(\frac{h+h_s}{10^5}\right)^{10}\exp\left(-\frac{h+h_s}{1000}\right)\right.$$ $$\left.+2.7\times10^{-16}\exp\left(-\frac{h+h_s}{1500}\right)\right]+C_n^2(h_0)\left(\frac{h_0}{h}\right)^p, \quad h>h_0,$$ where v is the high-latitude average wind speed (typically 21 m/s), h_s is the height of ground above sea level, h_0 is the height aboveground of scintillometer instruments (used to estimate C_n^2), $C_n^2(h_0)$ is the average refractive index structure function at h_0, and altitude varies from the reference height h_0 aboveground of the scintillometer instrumentations to the maximum height of the beam (or transmitter) aboveground. The parameter M characterizes an average value of the random background turbulence at altitudes generally above 1 km. The power law parameter p depends on the temporal hour of the day that measurements are made, which requires the local times at which sunrise and sunset occur.	
SLC [32, 33]	The submarine laser communication-day (SLC-day) model is frequently used to describe daytime turbulence conditions at inland sites and is given by $$C_n^2(h) = \begin{cases} 0 & 0m < h < 19m \\ 4.008\times10^{-13}\,h^{-1.054} & 19m < h < 230m \\ 1.300\times10^{-15} & 230m < h < 850m \\ 6.352\times10^{-7}\,h^{-2.966} & 850m < h < 7000m \\ 6.209\times10^{-16}\,h^{-0.6229} & 7000m < h < 20,000m \end{cases}$$	

Optical turbulence models	Model description and the physics involved	Applicability for FSO communication
Malaga turbulence models [34]	A general turbulence model with a unified PDF of the turbulence-induced fading; irradiance fluctuations of an unbounded optical wave front (plane and spherical waves) under all irradiance fluctuation conditions. Observed field at the receiver to consist of three terms: (1) LOS contribution, U_L; (2) quasi-forward scattered by the eddies on the propagation axis, U_S^C; and (3) due to energy scattered by the receiver by off-axis eddies (see [40] for details)	
Attenuation and scattering models [11, 35]	See Hemani, Kar, and Jain's paper free-space optical channel models, Eq. (2.85), PLUS Jennifer's paper beam extinction (absorption and scattering) due to aerosols, fog, rain, snow, and molecules: $$\beta_\nu = \beta_{sca,\nu}^{aer} + \beta_{abs,\nu}^{aer} + \beta_{sca,\nu}^{mol} + \beta_{abs,\nu}^{mol}$$ β_ν comprises the aerosol scattering, aerosol absorption, molecular scattering, and molecular absorption at the frequency, ν. $$r_0 = \left[0.423k^2 \sec(\theta) \int_{h_0}^{h_0+L} C_n^2(h)\, dh \right]^{-3/5}$$ $$r_0 = \left[0.423k^2 \sec(\theta) \int_{h_0}^{h_0+L} C_n^2(h) \left\{ \frac{L+h_0-h}{L} \right\}^{5/3} dh \right]^{-3/5}$$	

2.2.6   Atmospheric Communication Channels and Effects Applicable to 5G/6G and IoT Applications

2.2.6.1   Free-Space Optical Communication Technologies for 5G Evolution and 6G

FSO communication with the potential for establishing high-speed communications with capabilities of small size, weight, and power (SWaP) can support 5G networks for future industry and society, along with artificial intelligence (AI) and the Internet of Things (IoT). 5G is also expected to contribute to the upgrading of the multimedia communication services with its technical features such as high speed, high capacity, low latency, and massive connectivity. The mobile communication system equipped with all-optical devices will be the future 5G networks. For the future 6G network, the coverage extension will include non-terrestrial network to provide services for drones, flying cars, ships, and space stations since they are not covered by conventional cellular networks. Eventually, in order to consider super-extension technology for the 6G networks, the utilization of geostationary satellites (GEO), low Earth orbit satellites (LEO), and high-altitude platforms (HAPs) will need to be integrated as shown in the earlier Fig. 2.9.

Smartphones have been explosively popularized by high-speed wireless communication technology in which 5G developed around the spring of 2020 and mobile communication system can provide a basic technology to support future industry and society, along with artificial intelligence (AI) and the Internet of Things (IoT). With 5G, upgrading of the multimedia communication services achieving high speed, high capacity, low latency, and massive connectivity is possible. The sixth-generation (6G) technology with even more advanced technical features will support industry and society in the 2030s. 5G is the first-generation mobile communication system supporting high-speed wireless data communications of several Gbit/s. The high speed and high capability of 5G technology can even be further improved with FSO communications, which will still require improvements in the coverage and uplink performance in non-line-of-sight (NLOS) environments both of which need accurate understanding of the atmospheric effects on optical communications. The development of 6G from the 5G evolution will mostly consider the requirement of the wider and deeper diffusion to handle more advanced services, integration of multiple new cases using much more advanced signal processing techniques, and various high-speed devices. It is important to note that all these 5G and 6G evolutions using FSO communications will still require a thorough understanding and correct knowledge of the propagation of optical beams through the atmosphere which will depend on the propagation links (slant), altitudes of the optical transceivers' locations, and different scenarios of connecting the future optical towers with the Internet. Communication environments such as high-rise buildings, drones, moving and future flying cars, airplanes, and even space will all be natural activity areas. There will also be future needs for developing FSO communications at sea and under the sea.

Atmospheric effects of FSO communication will therefore need to extend the present 5G/6G coverage extension technology to include both terrestrial and non-terrestrial networks. Atmospheric propagation links need to be investigated three-dimensionally including both vertical and slant paths. Long-distance optical wireless transmission over several tens of kilometers needs to be considered because of the wireless backhaul and integrated access backhaul (IAB). FSO can be considered as a viable option to interconnect cell towers to create backhaul network to deliver 20 Gbit/s between cellular towers. Cell size is decreasing to a few hundred meters; FSO technology can connect such cells with high capacity in next-generation wireless networks with low-cost compared to fiber technology. Finally, to apply FSO communication to remote places, it will be necessary to include geostationary satellites (GEO), low Earth orbit satellites (LEO), and high-altitude platform (HAP) satellites. Since FSO signals are propagated through the air, limitations of achieving high-data-rate communication by atmospheric effects must be considered. Some of these limitations include physical barrier (e.g., obstruction), atmospheric turbulence, misalignment or pointing error, background noise (e.g., sunlight, diffused extended noise from the atmosphere), beam spreading and path loss, absorption and scattering loss (atmospheric and water molecules), and cloud coverage. Attenuation functions that model all the limitation factors over FSO links are significant for reliable FSO communication system performance.

Today, the needs and therefore the capabilities of wireless networks are progressing more rapidly than ever. The benefits of connecting end-to-end integrated network solutions providing reliable service for both fixed and mobile users are well established now. In order to connect high-capacity smart devices with increased number of broadband providers available today, there is tremendous increased demand for mobile users where free-space optical wireless communication (FSOC) is absolutely necessary. Optical wireless communication has the capability of solving high-capacity, high-data-rate transmissions and receptions. The 5G/6G network promises to deliver unprecedented data rates for mobile users with a very high bandwidth for a number of applications including Internet of Things (IoT). FSOC can play a very important role in *Beyond 5G and 6G Wireless Networks* for both short-range and long-range applications. FSO communication systems can provide high-speed mobile access in hotspots based on light fidelity (Li-Fi) technology. For long-range applications, FSO communication systems can provide both fixed and moving (with tracking and directed FSO) links such as between buildings/towers and in space, ground-to-space, and space-to ground. Atmospheric effects, however, put the limits for reliable and efficient links. For 5G/6G and IoT applications using optical communication, it is important to understand and analyze different propagation scenarios and investigate their effects accurately.

FSO technology for 5G networks and beyond: FSO is a cost-effective and wide-bandwidth solution as compared to traditional backhaul solutions including fiber backbone and backhaul. Because of atmospheric effects on FSO links, a key requirement for deploying an FSO backhaul network will be to increase the reliability of FSO systems for high-data-rate communications. The theoretical models proposed to enhance FSO links for 5G performance efficiency will be similar to those applied to near-ground and low-altitude atmospheric turbulence and scattering except that an integrated *mobile* FSO system needs to be considered. The models for fixed terminals can be used for indoor and outdoor setups.

2.2.7 Underwater Optical Communications

The aim of the free-space optical (FSO) wireless communication for global connectivity is to develop extreme coverage extension that can be used in all kinds of medium including the sky, sea, and space, which are not connected by FSO yet. FSO communication can also be expected to be applied to future applications such as flying cars and space travels. A general sketch of an integrated system architecture for space/aerial/terrestrial/underwater terminals showing site diversity for laser communication links is shown in Fig. 2.9 in the earlier section. Underwater communication terminals are shown in the figure for potential underwater free-space optical communication (uFSO).

Underwater free-space optics (uFSO) technology can establish communication links between autonomous underwater vehicles (AUVs) equipped with the optical transmitter and receiver as well as currently used remotely operated underwater vehicles (ROVs). The transmission distance of a few meters to a hundred of meters will depend on the wavelength of the light (usually in the blue-green range) and the type of seawater, and operating depth. Furthermore, the optical communication links can include both underwater and space FSO terminals to cover both water and space. In order to develop the uFSO technology for reliable data transfer, it is therefore important to investigate the propagation effects of optical beams through air (space) and water and at the air-water interface. Accurate channel models of optical propagation through water for absorption and scattering at the optical wavelengths exist [36], but the statistical characterization of the air-water interface is not much available in the literature. The author has contributed some research in this subject which is essential to characterize the statistical PDF of laser intensity fluctuations to be useful in calculating the FSO and uFSO communication system performance.

Underwater Free-Space Optical (uFSO) Communication Links with Gbit/s Data Rate

In order to include FSO underwater terminal to connect with air/space optical communication terminals, various underwater optical networks capable of high-transmission-speed, large-bandwidth underwater data communication links are needed to develop. Underwater FSO communication can be a part of the global communication networks to establish Internet connectivity in remote areas. The U.S. Defense Advanced Research Projects Agency (DARPA) is interested in developing a blue-spectrum submarine laser communications system which will be able to link submerged submarines with nearby aircraft [38]. Previously, DARPA developed a similar technology which tested blue-green laser to match with a receiver at 455.6 nm, whereas the downlink was a blue-green diode-pumped laser compatible with existing submarine receivers at 532 nm. The design and development of reliable uplink and downlink transceivers using an aircraft as a surrogate satellite/ medium- to high-altitude aircraft platform are important for ultimately the space/ air/sea optical networks for global Internet connectivity optical *anywhere anytime*. An underwater wireless communication technology with high data rate, transmission speed, and longer ranges (~100s of meters) will be essential for real-time video streaming and data upload. Recently, a laser diode-based underwater wireless communication system with a data rate of 1.5 Gbit/s over a 20 m channel using NRZ-OOK modulation scheme has been reported [37, 43].

A brief summary of some of the recent essential results of underwater-related FSO communications is given below:

Recently, there has been increasing interest in underwater wireless media for underwater communications. Because of the limitations of low bandwidth and low data rate, underwater free-space optics (uFSO) is gaining more attention to provide high-speed underwater optical communication at low latencies but for a limited communication range. Some of the basic research in the propagation and the properties of blue-green (450–550 nm) optical wavelength regions are reported [40, 41]. Laser diodes (LDs) were proposed [42] in order to increase the data rate compared to the light-emitting diodes (LEDs), which have insufficient bandwidth and low transmission distance. Some of the other researches for uFSO communications have been reported [43] of achieving a data rate of 4.88 Gbit/s by using quadrature amplitude modulation-orthogonal frequency division multiplexing (QAM-OFDM).

2.2.7.1   Channel Models of uFSO Systems/Underwater Propagation Characteristics of Optical Wavelength

Some of the researchers have considered vector radiative transfer theory, variable water composition, and inherent properties of light. When light beams propagate through water, it suffers from *absorption* and *scattering* which again depends on the types of oceanic water such as (1) pure sea water, (2) coastal ocean water, and (3) turbid harbor water. There is also *oceanic turbulence* effect, which can also limit the optical propagation and available data rate. Scintillation by water turbulence is due to rapid variations in the refractive index generated by fluctuations in the water medium parameters like pressure, density, salinity, and temperature. Using the typical atmospheric turbulence effects on FSO propagation and transmission, uFSO channels have been reported using the Kolmogorov model for intensity fluctuations of received light for underwater turbulence model [44].

Impulse response: The channel impulse response of underwater optical channel has been discussed in [47] under different water conditions like link distance and transmitter/receiver parameters. Their results demonstrated that the channel delay dispersion is negligible in most practical cases and therefore does not suffer from any intersymbol interference (IS) which does not require complex channel equalization at the receiver.

2.2.7.2 Air-Water Interface Statistical Model (Applicable for Satellite (or Airborne)-to-Underwater or Underwater Terminal Such as Submarine to a Satellite/Airborne Terminal)

Random air-water interface has been very complex until recently as the author and his colleagues have reported [46]. This subject is discussed below. To design and develop a uFSO communication system, an accurate modeling of pointing errors and misalignment modeling are necessary to evaluate the uFSO communication system performance under misalignment condition. Point spread function (PSF) of a laser source in spatial coordinates and spatial frequency domain are useful parameters to determine the performance loss due to misalignment. Impacts of transmitter parameters like divergence and elevation angles due to jitter of transceivers caused by random water surface can be determined from the random PSF reported [45]. This topic is also discussed below. An excellent comprehensive survey paper [39] discusses the details of underwater optical wireless networks covering different aspects of cutting-edge underwater optical wireless communications from a layer-by-layer perspective.

Optical propagation through the random wavy air-water interface relevant to underwater optical communications is discussed [46]. Their paper is important to understand the physics of optical wave interactions through random air-water interface and to quantify the beam wander effects for both reflection and transmission cases which then can be used for estimating the underwater free-space optical (uFSO) communication link performance to apply to various communication scenarios such as uplink and downlink for underwater and aerial/space terminals. Using a statistical model for rough water surface discussed in [45], Gram-Charlier model for wave-angle statistics, and finite-element ray approximation of the incident wave front, the authors computed the irradiance distribution in a viewing plane. When a transmitting laser beam propagates through air, air-water interface, and then through water, there is a significant distortion due to high geometric phase aberrations introduced by the random motion of water surface wave. A significant reduction in the received communications signal can limit the data transfer capability and available data rates. Beam wander effects are caused by random angular displacements of the centroid of the fluctuation laser beam. The details of the derivations of the probabilistic models are described elsewhere [46].

The essential results are the following:

Case 1: Reflection Case

Figure 2.10 shows the angles of interest for reflection of a laser beam by random air-water interface where the angles are defined as [46]

where

θ_I = angle of incidence for the mean or calm surface
θ_S = slope of the time-varying surface at the moment of contact

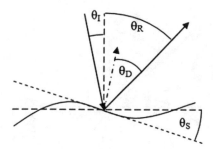

Fig. 2.10 The angles of interest for reflection of a laser beam by random air-water interface where the angles are defined as [46]

θ_R = angle of the reflected beam
θ_D = angle by which this reflection deviates from its position in the case of a calm, flat surface

If some form of probability density function (PDF) for the wave slopes, $PDF_S(\theta_S)$, is known, the PDF of the angular fluctuations of an emergent angle (reflection case) is given by

$$PDF_D(\theta_D) = \frac{1}{2} PDF_S\left(\frac{\theta_D}{2}\right) \qquad (2.15)$$

Case 2: Transmission Case

Figure 2.11 shows the transmission case angles of interest.
   For transmission through the random air-water interface, the Snell's law has been applied:

$$\frac{\sin(\theta_I + \theta_S)}{\sin(\theta_T + \theta_S)} = \frac{n_2}{n_1}$$

For air and water under common condition, the above ratio is 4/3. Further using the approximations, the following equations are obtained:

$$\theta_T = \arcsin\left(\frac{3}{4}\sin(\theta_I + \theta_S)\right) - \theta_S$$

$$\theta_D = \theta_T - \theta_T|_{\theta S=0}$$

$$= \arcsin\left(\frac{3}{4}\sin(\theta_I + \theta_S)\right) - \theta_S - \arcsin\left(\frac{3}{4}\sin(\theta_I)\right)$$

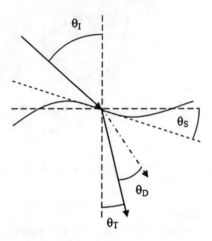

Fig. 2.11 The transmission case angles of interest

This equation is not monotonic in the inverse, and there are multiple solutions when solving for θ_S.

Using the simplest estimate for the beam wander PDF, the PDF is estimated as

$$PDF_D\left(\theta_D\right) \cong 4PDF_S\left(-4\theta_D\right)$$

$$\theta_D \cong -\frac{1}{4}\theta_S + O(2)$$

$$PDF_D\left(\theta_D\right) \cong 4PDF_S\left(-4\theta_D\right) \tag{2.16}$$

Using these above results, the beam wander effects applicable to both reflection and transmission cases and the uFSO communication parameters such as BER as a function of SNR for both reflected and transmitted are shown in the report [46].

2.2.7.3 Adaptive Optics for Underwater Communication: Mitigation of Laser Propagation Through Random Air-Water Interface

A new concept for mitigating signal distortions caused by random air-water interface using an adaptive optics (AO) is reported by the author and his colleague [49]. The authors demonstrated the feasibility of correcting the distortions using AO in a laboratory water tank for investigating the propagation effects of a laser beam through air-water interface. A typical adaptive optics system consists of a fast steering mirror (FSM), deformable mirror (DM), and a Shack-Hartmann wavefront sensor (WFS). Results are presented for a number of metrics including Strehl ratio,

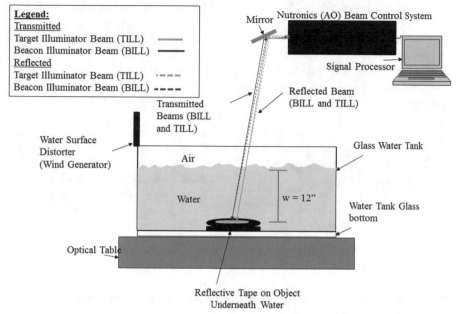

Fig. 2.12 Water tank experiment with adaptive optics (AO) beam control system. (*Reprinted with permission from Proc. SPIE, 2021* [49])

which quantifies the quality of optical image formation from diffraction-limited optical system and is relevant to the received signal for FSO communications operating through air and water terminals. The AO system includes an integrated beacon illuminator (BILL) and track illuminator (TILL) designed to compensate weak random fluctuations of optical beam due to random air-water interface and is shown in Fig. 2.12.

The TILL and BILL propagate through air-water interface and water to an object under water with a reflective tape, and the beam returns from the reflective tape back to the AO system. A fan at various speeds generated water distortions. Strehl ratio measured quantifies the capability of AO in correcting the distortions, and the results are shown in Fig. 2.13. The Strehl ratio is a power-in-the-bucket metric for both AO off ("open loop") and AO on ("closed loop") conditions for a fan speed of 12.7 mph (5.68 m/s) wind to generate water distortions. The results clearly demonstrate an improvement of the beam quality by AO. The AO technology concept for mitigating atmospheric turbulence-induced or random air-water interface is a very useful technique in order to improve the FSO communications performance. The typical application for improving FSO communications performance can be from satellite/UAV/aircraft/HAP to underwater terminal link or from underwater to satellite/UAV/aircraft/HAP terminal link.

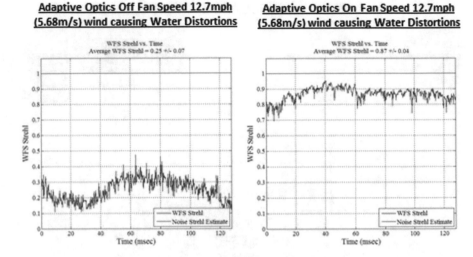

Fig. 2.13 Adaptive optics (AO) improvement of the Strehl ratio. (*Reprinted with permission from Proc. SPIE, 2021* [49])

2.2.8 Non-isotropic (Non-Kolmogorov) Atmospheric Model: Relevant to FSO

For the last few years, free-space optical (FSO) has increasingly attracted for a variety of applications, such as uplinks and downlinks between ground platforms and space platforms and aircraft and ships, developing of the future 5G/6G technology, and among mobile or stationary terminals to solve the *last-mile* problem. High-bandwidth wireless communications links for these applications require optical propagation through atmospheric turbulence where the laser/optical signals for both transmitting and receiving beams suffer intensity fluctuations called *scintillation*. In addition, beam spreading and beam wander caused by atmospheric turbulence result in power loss and degrade the FSO communication systems by decreasing the received SNR and increasing bit error rate (BER) due to fading statistics. The physics of the origins of atmospheric turbulence is clearly discussed and explained by the authors in Refs. [6, 11]. Commonly accepted turbulence strength parameter, C_n^2, defines equating structure function and statistical ensembled random atmospheric refractive index fluctuations between two points. The turbulence strength parameter C_n^2 depends on a number of factors including temperature, humidity, and wind speed at a location. The previous section discussed a number of recently accepted C_n^2 profiles as a function of altitude. Most of the scintillation theories assume the turbulence to be isotropic and are reasonably valid for weak to somewhat strong turbulence situations. For FSO communications involving long-distance ground-ground horizontal/terrestrial links as well as ground-satellite links along slant path, the propagation path can easily include strong turbulence regimes under severe weather conditions. For designing and developing highly reliable FSO communications system, accurate values of the randomly varying turbulence-related parameters are necessary. Most of the scintillation theories for implementing FSO communications systems are based on Kolmogorov and isotropic model of turbulence. Because of the demands of developing very high data rates in 100s of Gbit/s to operate under long distance and strong turbulent paths, it is important to investigate the non-isotropic and non-Kolmogorov statistical models and their applicability in FSO communications system design. Some of the results for the impacts of non-isotropic and non-Kolmogorov turbulence models on FSOC communications are summarized below [48].

2.2.8.1 Anisotropic Power Spectrum (Scale Dependent)

As mentioned earlier, in real-world situations atmospheric turbulence does not always follow the Kolmogorov power spectrum model since the velocity and temperature fluctuations might not obey the same law from which the Kolmogorov's theory was developed which are generally used for isotropic turbulence. The relationship between temperature structure function and derived power spectrum is more complex. Authors in a recent paper extended the power spectrum to be valid for scale-dependent anisotropic turbulence. For evaluating FSO communications parameters such as SNR and BER for long horizontal path at high altitude or slant path from ground-to-satellite scenarios, the transmitted and received beam statistics might depend on anisotropic nature of turbulence.

2.2.8.2 The Atmospheric Channel Characteristics for Anisotropic Turbulence

The structure function and power spectrum are the main quantities to analyze the effects of atmospheric turbulence due to random spatial and temporal variations of atmospheric refractive index. For non-Kolmogorov turbulence, the structure function can be written as [48]

$$D_n(R) = \tilde{C}_n^2 R^{\alpha-3} = \tilde{C}_n^2 \left[\frac{x^2}{\varsigma_{x\_\text{eff}}^2} + \frac{y^2}{\varsigma_{y\_\text{eff}}^2} + z^2 \right]^{\frac{\alpha-3}{2}} \tag{2.17}$$

$$l_{0x} \ll x \ll L_{0x}, l_{0y} \ll y \ll L_{0y},$$

and the corresponding power spectrum is

$$\Phi_n(\kappa,\alpha) = \frac{A(\alpha) \cdot \tilde{C}_n^2 \cdot \zeta_{x\_\text{eff}} \cdot \zeta_{y\_\text{eff}}}{\left(\zeta_{x\_\text{eff}}^2 \cdot \kappa_x^2 + \zeta_{y\_\text{eff}}^2 \cdot \kappa_y^2 + \kappa_z^2\right)^{\frac{\alpha}{2}}}$$

$$3 < \alpha < 4, \quad \kappa_x, \kappa_y, \kappa_z > 0,$$

(2.18)

where α is the index for the power law and β is a dimensional constant.

$\tilde{C}_n^2 = \beta \cdot C_n^2$ is the generalized turbulence structure parameter, and the two unitless effective anisotropic factors are ζ_{x_eff} and ζ_{y_eff} along x and y directions, respectively. $A(\alpha)$ in the above equation is given by

$$A(\alpha) = \frac{\Gamma(\alpha-1)}{4\pi^2} \cos\left(\alpha \frac{\pi}{2}\right), \quad 3 < \alpha < 4,$$

$\Gamma(x)$ is a gamma function.

To analyze and evaluate the FSO communications performance through atmospheric turbulence, first the log-amplitude fluctuation variance or equivalent intensity fluctuation variance σ_I^2 or simply the scintillation index for the turbulence model at the communication wavelength, λ, or the wave number $\kappa = 2\pi/\lambda$ and the propagation path length L are required. The scintillation index is then evaluated from an integral equation [4] involving the power spectra defined above. Note that the atmospheric channel models, $C_n^2(h)$, as outlined earlier are appropriate for isotropic turbulence Kolmogorov models. To extend the channel models for non-isotropic turbulence non-Kolmogorov model, $C_n^2(h)$ for non-isotropic turbulence should be used as follows: non-isotropic/non-Kolmogorov turbulence channel model,

$$\tilde{C}_n^2(h) = \beta C_n^2(h)$$

(2.19)

where β is the constant defined earlier, and $C_n^2(h)$ is the conventional Kolmogorov turbulence channel model.

Anisotropy introduces a *rescaling* of turbulence, which can impact FSO propagation through turbulence specially for long distance and strong turbulence path, and therefore can limit the FSO communications performance in terms of availability of maximum data rate and communication link. The FSO communications link scenarios can include ground-to-aircraft/HAP/satellite platforms as well as data transfer between satellites. Accurate turbulence models can improve the FSO performance by designing the system properly avoiding extremely costly improvements and failure rates.

2.3 Summary

This chapter discusses the fundamental physics of space optical propagation applicable to integrated space/aerial, terrestrial, and underwater communication links. Slant-path free-space optical (FSO) communication links between an optical ground station (OGS) and the satellite at various orbits and altitudes include GEO, LEO, HAP, and small or even microsatellites that are analyzed with appropriate accepted optical channel models and atmospheric propagation parameters. Various FSO transmitting beam shapes to optimize long-distance data transmission are discussed. FSO communication technologies for 5G networks and future 6G evolution are explained. Underwater free-space optical (uFSO) communications for potential global connectivity at Gbit/s data rate is presented.

References

1. Arun K. Majumdar and Jennifer C. Ricklin, *Free-Space Laser Communications; Principles and Advances,* Springer, New York, NY, 2008
2. Arun K. Majumdar, Chapter 2, *Advanced Free Space Optics (FSO): A Systems Approach,* Springer Science+Business Media, New York 2015.
3. Arun K. Majumdar, *Optical Wireless Communications for Broadband Global Internet Connectivity: Fundamentals and Potential Applications,* Elsevier, Amsterdam, Netherlands 2019.
4. Larry C. Andrews and Ronald L. Phillips, Laser Beam Propagation through Random Media, Second Edition, SPIE PRESS, Bellingham, Washington 2005.
5. Lake A. Singh, William R. Whittecar, Marc D. DiPrinzio, Jonathan D. Herman, Matthew P. Ferringer & Patrick M. Reed, Low cost satellite constellations for nearly continuous global coverage, Nature Communications, (2020) 11:200 (2020) 11:200 | https://doi.org/10.1038/s41467-019-13865-0| www.nature.com/naturecommunications
6. V. I. Tatarskii, *The Effects of the Turbulent Atmosphere on Wave Propagation,* Translated from Russian by Israel Program for Scientific Translations Ltd, IPST Cat. No. 5319, UDC 551.510, ISBN 0 7065 0680 4, available from the U.S. Department of Commerce, NTIS, Springfield, VA 22151 TT-68-50464, 1971.
7. J. W. Strohbehn, ed., *Laser Beam Propagation in the Atmosphere,* Springer-Verlag, Berlin, 1978.
8. A. M. Prokhorov, F. V. Bunkin, K. S. Gochelashvily, and V. I. Shishov, Laser Propagation in turbulent media, Proc. IEEE.63, 790-811 (21975).
9. D. L. Fried, Scintillation of a ground-to-space laser illuminator, J. Opt.Soc. Am 57, 980-983 (1967).
10. P. O. Minott, Scintillation in an earth-to-space propagation path, J. Opt.Soc. Am 62, 885-888 (1972).
11. Jennifer C. Ricklin, Stephen M Hammel, Frank D. Eaton, Sventa l. Lachinova, Atmospheric channel effects on Free-Sapce Laser Communication, Chapter2 of the Book, Arun K. Majumdar and Jennifer C. Ricklin, *Free-Space Laser Communications; Principles and Advances,* Springer, New York, NY, 2008.
12. Hemani Kaushal1 and Georges Kaddoum, Optical Communication in Space: Challenges and Mitigation Techniques, DOI https://doi.org/10.1109/COMST.2016.2603518, IEEE Communications Surveys & Tutorials, 2016.

13. Shuhui Li and Jian Wang, Adaptive free-space optical communications through turbulence using self-healing Bessel beams, SCIENTIFIC REPORTS, 7:43233 | DOI: https://doi.org/10.1038/srep43233, 23 February 2017 www.Nature.com/scientificreports
14. Nokwazi Mphuthi, Lucas Gailele, Igor Litvin, Angela Dudley, Roelf Botha, and Andrew Forbes, Free-space optical communication link with shape-invariant orbital angular momentum Bessel beams, Applied Optics Vol.58, Issue 16, pp.4258-4264 (2019).
15. PHILIP BIRCH, INIABASI ITUEN, RUPERT YOUNG, AND CHRIS CHATWIN, Long Distance Bessel Beam Propagation Through Kolmogorov Turbulence, Journal of the Optical Society of America A, September 21, 2015.
16. Minghao Wang1,2, Xiuhua Yuan1, Peng Deng2, Wei Yao3, Timothy Kane2, Application of Aperture Truncated Airy Beams in Free Space Optical Communications, SPACOMM 2018 : The Tenth International Conference on Advances in Satellite and Space Communications, ISBN: 978-1-61208-624-8, IARIA 2018.
17. *Nathaniel A. Ferlic, Miranda van Iersel, Daniel A. Paulson, Christopher C. Davis*, Propagation of Laguerre-Gaussian and I_m Bessel beams through atmospheric turbulence: A computational study, Proceedings Volume 11506, Laser Communication and Propagation through the Atmosphere and Oceans IX; 115060H (2020) https://doi.org/10.1117/12.2567348, SPIE 22 August 2020.
18. Ing. Peter Barcík, Doctoral Thesis, OPTIMAL INTENSITY DISTRIBUTION IN A LASER BEAM FOR FSO COMMUNICATIONS, BRNO University of Technology, 2016
19. Ruediger Grunwald & Martin Bock (2020) Needle beams: a review, Advances in Physics: X, 5:1, 1736950, DOI: https://doi.org/10.1080/23746149.2020.1736950, ADVANCES IN PHYSICS, Vol.5, No. 1, 1736950, Taylor & Francis, 2020.
20. Kaiwei Wang, Lijiang Zeng, Chunyong Yin, Influence of the incident wave-front on intensity distribution of the non-diffracting beam used in large-scale measurement, Optics Communications 216 (2003) 99-103.
21. V. Kollárová, T. Medřik, R. Čelechovský, Z. Bouchal, O. Wilfert, Z. Kolka, Application of non-diffracting beams to wireless optical communications. t: https://www.researchgate.net/publication/336453975
22. Ömer F. Sayan a, Hamza Gerçekcioğlu a, Yahya Baykal, Hermite Gaussian beam Scintillations in weak atmospheric turbulence for aerial vehicle laser communications, Optics Communications, October 2019, DOI: https://doi.org/10.1016/j.optcom.2019.12473.
23. Hyo-Chang Kim, Yeon H. Lee, Hermite-Gaussian and Laguerre-Gaussian beams beyond the paraxial approximation, Optics Communications, Vol. 169, Issues 1-6, Pages 9-16, 1 October 1999.
24. Mehmet Yuceer and Halil T. Eyyuboglu, Laguerre-Gaussian beam scintillation on slant paths, Appl. Phys. B (2012) 109:311–316 DOI https://doi.org/10.1007/s00340-012-5186-3.
25. Stamatios V. Kartalopoulos, "Free Space Optical Networks for Ultra-Broad Band Services," IEEE/Wiley Publication, Hoboken, New Jersey, 2011.
26. Y. Han Oh, J. C. Ricklin, E. Oh, S. Doss-Hammel and F. D. Eaton, "Estimating optical turbulence effects on free-space laser communication: modeling and measurements at ARL's A_LOT facility," SPIE Vol. 5550, 247-255 (2004).
27. R. W. Smith, J. C. Ricklin, K. E. Cranston and J. P. Cruncleton, "Comparison of a model describing propagation through optical turbulence (PROTURB) with field data," SPIE Vol. 2222, 780-789 (1994).
28. R. E. Hufnagel, "Variations of atmospheric turbulence," Tech. Report, 1974.
29. G. C. Valley, "Isoplanatic degradation of tilt correction and short-term imaging systems," Appl. Opt., vol. 19, no. 4, pp. 574–577, February 1980.
30. R. R. Beland, "Propagation through atmospheric optical turbulence," in The Infrared and Electro-Optical Systems Handbook, F. G. Smith, ed. (SPIE Opt. Eng. Press, Bellingham, 1993), Vol 2, Chap. 2.
31. L. C. Andrews, R. L. Phillips, R. Crabbs, D. Wayne, T. Leclerc, and P. Sauer, "Atmospheric channel characterization for ORCA testing at NTTR," Proc. SPIE 7588 (2010).

32. M. G. Miller and P. L. Zieske, *Turbulence environmental characterization* (Rome Air Development Center, Griffiss Air Force Base, N.Y., RADC-TR-79-131, 1979).
33. R. R. Parenti and R. J. Sasiela, "Laser-guide-star systems for astronomical applications," J. Opt. Soc. Am. A **11**(1), 288-309 (1994).
34. A. Jurado-Navas, J. M. Garrido-Balsells, J. F. Paris, and A. Puerta-Notario, "A unifying statistical model for atmospheric optical scintillation," arXiv preprint arXiv:1102.1915, 2011.
35. Hemani Kaushal, Subrat Kar and Vk Jain, Free-Space Optical Channel Models, Chapter 2, H. Kaushal et al., Free Space Optical Communication, Optical Networks, DOI https://doi.org/10.1007/978-81-322-3691-7_2, Springer (India) Pvt. Ltd. 2017, January 2017.
36. Hassan M. Oubei, Chao Shen, Abla Kammoun, Emna Zedini, Ki-Hong Park, Xiaobin Sun, Guangyu Liu, Chun Hong Kang, Tien Khee Ng, Mohamed-Slim Alouini and Boon S. Ooi, Light based underwater wireless communications, *Jpn. J. Appl. Phys.* 57 08PA06, 17 July 2018.
37. CHAO SHEN, YUJIAN GUO, HASSAN M. OUBEI, TIEN KHEE NG, GUANGYU LIU, KI-HONG PARK, KANG-TING HO, MOHAMED-SLIM ALOUINI, AND BOON S. OOI1,* 20-meter underwater wireless optical communication link with 1.5 Gbps data rate, Optics Express, Vol. 24, No. 22 | 31 Oct 2016 | OPTICS EXPRESS 25502, 24 October 2016.
38. Military & Aerospace Electronics, Jan 31st. 2010. https://www.militaryaerospace.com/home/article/16723525/darpa-pushes-submarine-laser-communications-technology-for-asw-operations
39. Nasir Saeed, Abdulkadir Celik, Tareq Y. Al-Naffouri, Mohamed-Slim Alouini, Underwater Optical Wireless Communications, Networking, and Localization: A Survey, arXiv:1803.02442v1 [cs.NI] 28 Feb 2018, May 2019, Ad Hoc Networks 94:101935, DOI: https://doi.org/10.1016/j.adhoc.2019.101935
40. S. Q. Duntley, "Light in the sea∗," J. Opt. Soc. Am., vol. 53, no. 2, pp. 214–233, Feb. 1963.
41. G.D. Gilbert, T.R. Stoner, and J.L. Jernigan, "Underwater experiments on the polarization, coherence, and scattering properties of a pulsed blue-green laser," Proc. SPIE, vol. 0007, pp. 07 – 14, Jun. 1966.
42. F. Hanson and S. Radic, "High bandwidth underwater optical communication," Appl. Opt., vol. 47, no. 2, pp. 277–283, Jan. 2008.
43. J. Xu, Y. Song, X. Yu, A. Lin, M. Kong, J. Han, and N. Deng, "Underwater wireless transmission of high-speed QAM-OFDM signals using a compact red-light laser," Opt. Express, vol. 24, no. 8, pp. 8097– 8109, Apr. 2016.
44. F. Hanson and M. Lasher, "Effects of underwater turbulence on laser beam propagation and coupling into single-mode optical fiber," Appl. Opt., vol. 49, no. 16, pp. 3224–3230, Jun. 2010.
45. William C. Brown and Arun K. Majumdar, Point-spread function associated with underwater imaging through a wavy air-water interface: theory and laboratory tank experiment, APPLIED OPTICS, Vol. 31, No. 36, 20 December 1992.
46. Arun K. Majumdar, John Siegenthaler, Phillip Land, "Analysis of optical communications through the random air-water interface: feasibility for underwater communications," Proc. SPIE 8517, Laser Communication and Propagation through the Atmosphere and Oceans, 85170T (24 October 2012); doi: https://doi.org/10.1117/12.928999.
47. Chadi GABRIEL, Mohammad-Ali KHALIGHI, Salah BOURENNANE, Pierre LEON, Vincent RIGAUD, Channel Modeling for Underwater Optical Communication, Conferences, 2011 IEEE GLOBECOM Workshops, DOI:https://doi.org/10.1109/GLOCOMW.2011.6162571. Corpus ID: 206692769
48. Italo Toselli and Olga Korotkova, General scale-dependent anisotropic turbulence and its impact on free space optical communication system performance, JOSA A, Vol. 32, No. 6, pp. 1017-1025/June 2015.
49. Phillip Land and Arun K. Majumdar, Demonstration of Adaptive Optics for mitigating laser propagation through a random air-water interface, Proc. SPIE, Volume 9827, Ocean Sensing and Monitoring VIII: 982703, 17 May 2016.

Chapter 3
Laser Satellite Communications: Fundamentals, Systems, Technologies, and Applications

3.1   Introduction

It is obvious that satellites play a crucial role in improving today's digital world dealing with almost every aspect of high-quality multimedia service and in developing more efficient access networks addressed by a number of technologies to meet the end user's communication needs. Satellite communications have evolved to serve the basic telecommunication needs of a majority of countries around the world by reaching directly to the home from space. Previously, telecommunications were limited by point-to-point communication which was accomplished either by cables (including fiber optic) and line of sight with limited ranges. Satellite communication has opened up a new space activity to home, business, and government users worldwide. Today, satellites are incorporating the multimedia information and personal communication. Satellite communications are also absolutely necessary for emergency preparedness and response, and disaster relief. Optical wireless communication, which has a number of advantages from data handling, data relaying, structuring, and implementation of complex optical networks, can offer all such possible solutions when integrated in an optical satellite. This section will describe the fundamentals of satellite communication in general first and then will address and emphasize on laser satellite communication. In order to design an efficient optical satellite for communication, a thorough systems design and characterization are therefore absolutely necessary. This will involve proper selections of transmitting laser and optical receiver subsystems. For example, extremely small laser-transmitting beam width and reduced-noise optical sources and detectors with high sensitivity will be needed for achieving high-bit-rate and high-capacity communication. Free-space optical (FSO) communication will include the application of deep-space communications between another planet and the Earth. Just recently, the rover *Perseverance* safely landed on Mars on February 18, 2021, for evidence of past life on Mars. Human missions to Mars will use lasers to communicate with Earth—NASA (reference: https://www.inverse.com/science/human-missions-to-mars-will-use-lasers-to-communicate-with-earth-nasa/amp).

© Springer Nature Switzerland AG 2022   63
A. K. Majumdar, *Laser Communication with Constellation Satellites, UAVs, HAPs and Balloons*, https://doi.org/10.1007/978-3-031-03972-0_3

3.2   New Network Characteristics

Based on the above advantages of satellite communication, it is obvious that satellite communication is absolutely necessary to provide ubiquitous connectivity with unconstrained geographical circumstances. Satellite optical communication has the technology potential of supporting a seamless global coverage of various locations such as land, sea, air, and space for establishing ubiquitous connectivity. This is also evident from the current developments of 5G and future 6G with the full support of artificial intelligence (AI) to be implemented in future between 2027 and 2030. To provide an FSO-based system using *all-optical* technology to develop a constant global Internet connectivity, it is expected to integrate terrestrial and satellite systems. Ideally, a single optical wireless system will be needed by integrating terrestrial, satellite, and airborne networks. Figure 3.1 shows a general architecture of integrated satellite/aerial networks.

3.2.1   Background

Free-space laser communications using atmosphere (space) as a channel was explored theoretically in the mid-1950s by Fried [5] using weak fluctuation propagation theory and a collimated Gaussian beam model. Results from the experiment using a continuous-wave (CW) argon laser transmitting at 0.488 µm and beamed at a GEOS-II satellite orbiting at approximately 1200 km were also reported [6]. A number of research

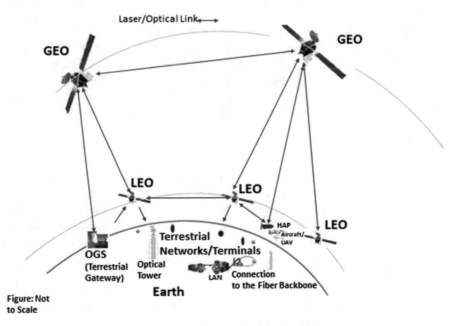

Fig. 3.1 A general architecture of integrated satellite/aerial networks. (Figure: Not to scale)

reports were published since then covering both theoretical aspects of optical propagation and uplink and downlink experimental results including ground-based lasers aimed at GEOS-III satellite (orbiting at 1800 and 2000 km) and with other various lasers to determine the parameters of interest to FSO communication. Later on, the Strategic Defense Initiative Organization (SDIO) Relay Mirror Experiment (RME) used projection from the ground sites at AMOS in Hawaii and both retroreflected and reflected laser beams from RME satellite at 350 km. Many excellent research papers, both theoretical and experimental, were later published to establish the feasibility of FSO communication between ground and space terminals. Some of the research by Andrews et al. [7] includes estimating communication parameters for both uplink and downlink propagation channels for a laser satellite communication system in GEO application of adaptive optics in ground-to-space communication to mitigate the atmospheric effects, which was reported by Tyson [8] in 1996.

The technology of FSO communications is now a mature technology to meet a growing demand for operational multi-Gbit/s data rate systems between ground and satellites/aircraft. This is being possible because of the recent technological advancement in space-based transceiver instruments and compact, fast optical devices for transmitting, receiving, and data processing either on board or at ground stations. The goal is to establish *all-optical* technology *anytime, anywhere*. As shown in Fig. 3.1, the global optical networks will include ground-to-satellite/aerial/HAP, satellite/aerial/HAP-to-ground, satellite/aerial/HAP-to-satellite/aerial/HAP, terrestrial, and satellite/aerial/HAP-to-underwater terminals. The future technology concept of global FSO communication will be aimed at satisfying the increasing high demands of high data rate and large communication capacity from all over the world including any remote locations. Since atmospheric effects on optical propagation ultimately put limits on available data rates and communication capacity, there are technical challenges, and potential ways to overcome them will be needed to design a space-based FSO communication systems to operate in the presence of atmospheric turbulence and scattering media. FSO communication is obviously useful in many application areas including 5G/6G telecom technologies.

3.2.2   Satellite Orbits: Various Types of Satellites

Laser satellites for communication move on various orbits around the Earth, which depends on the orbit altitudes from the Earth (https://www.esa.int/Enabling_Support/Space_Transportation/Types_of_orbits).

Types of orbit will be determined based on the orbit that would be the best for a satellite to use for satisfying specific communication goal. Some of the orbits commonly used are:

- Geostationary orbit (GEO)
- Medium Earth orbit (MEO)

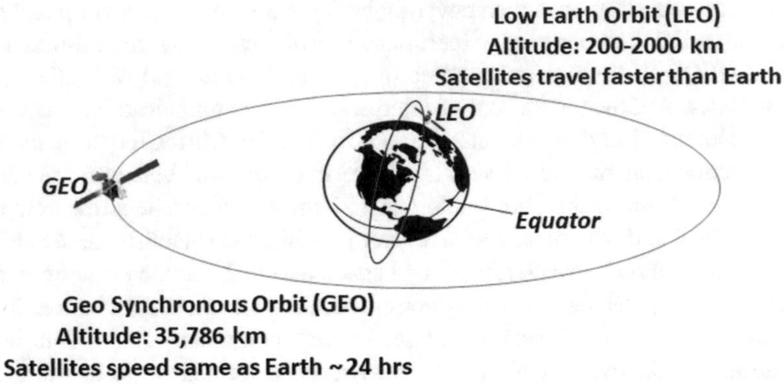

Fig. 3.2 GEO and LEO orbits for laser satellite communication

- Low Earth orbit (LEO)
- Polar orbit and Sun-synchronous orbit (SSO)
- Transfer orbits and geostationary transfer orbit (GTO)

Understanding of orbits dates back to Kepler in the seventeenth-century communication satellites, which were launched to many types of orbits. Some basic orbits commonly used in satellite communication systems include the low Earth orbit (LEO) and the geostationary Earth orbit (GEO). An orbit is the curve path that an object in space, such as a satellite, takes around another object, like the Earth due to gravity. Satellites are launched by rockets, which put them into orbit around the Earth, and the gravity keeps the satellite on its required orbit. Figure 3.2 shows GEO and LEO orbits commonly used for satellite communication systems.

LEOs are used for communications satellites and are usually at altitudes of several hundred to several thousand kilometers but could be as low as 160 km above Earth. Most commercial airplanes do not fly at altitudes much greater than ~14 km. NASA and military use LEO satellites, the International Space Station (ISS) is a LEO satellite, and the Space Shuttle flies in a LEO. A LEO has a period of ~90 min.

The *geostationary Earth orbit (GEO)* satellite is used for communication relay satellites. When the inclination of a GEO orbit with the equator is not zero, but the satellite motion is synchronized with the Earth's 24-h period, the orbit is termed as *geosynchronous orbit*. Russian satellites use a special elliptical orbit called Molniya orbit to provide northern polar coverage, which is a 12-h orbit inclined about 65° to the equator, but it may take at least three or four satellites in this orbit to provide continuous service to Russia. Figure 3.3 shows Molniya orbit.

In Chap. 1, the advantages of laser over RF communication relevant to satellite communications were discussed and explained. Comparison between RF and laser footprints of laser beam and RF beam from GEO and LEO orbit satellites was also discussed. It was pointed out that the primary benefit of laser communication arises from the narrow beam divergence of the laser.

Fig. 3.3 Molniya orbit: usually the period from "perigee + 2 h" to "perigee + 10 h" is used to transmit to the northern hemisphere

Some of the benefits of laser communications using the fundamental properties of laser beam propagation can be applied to increase communications capacity by delivering much higher data rate than existing RF communications with much less size, weight, and power (SWP) consumption. Optical communication spectrum is unregulated and license free, and commercially available optical transceivers in low Earth orbit to high-orbit FSO links are now being operational with Gbit/s speeds. Underwater laser communication is also adding to the overall communication architecture of satellite-to-underwater links. The technology concepts developed in this book for establishing *all-optical* technology will provide practical solution for broadband global Internet connectivity to remote locations.

3.3 Theoretical Background and Analysis

This section will discuss the fundamentals of optical propagation theory relevant to laser satellite communication.

3.3.1 Summary of Essential Results of Optical Propagation Relevant to FSO Communications System Analysis and Design

The potential applications of free-space optical/laser communications for satellite communication channels for high-data-rate transmission and reception have been around for the last three or more decades. There is some excellent published research already available in the literature [1–3]. The atmospheric channel between a ground terminal and a satellite/aerial terminal is in general random, and therefore the optical field originating from an optical transmitter and reaching the receiver undergoes various temporal and spatial changes in the communication signal. Both the transmitted and the received signals are therefore distorted due to the effects of atmospheric turbulence and scattering media. An FSO communication system designer needs to know the communication parameters based on the system requirements and the cost of the project. Tatarskii [4] and some other researchers laid some of the foundations of developing the theory of optical propagation through turbulent atmosphere, which can be applied to evaluate the FSO communication system parameters. The solution of optical wave equation provides the second- and fourth-order moment of the random optical field from which useful parameters such as mean irradiance, scintillation index, and covariance functions can be deduced. To analyze system performance and then to determine the communication design parameters, the main contributions for degradation of the signal quality due to beam wander and beam spreading, angle-of-arrival fluctuations, correlation length, and loss of spatial coherence are important. Finally, FSO communication performance metrics such as bit error rate (BER), received signal-to-noise ratio (SNR), and fading probability can be evaluated from the scintillation statistics, intensity probability density functions (PDFs), and appropriate mitigation techniques (such as aperture averaging, coding, adaptive optics) used. *Summary of results*, already discussed with detailed derivations in the literature [1–4], is the following:

3.3.1.1   Basic Parameters for Optical Propagation Through Atmospheric Turbulence

With the index of refraction at a point $\vec{r}$ in space, $n(\vec{r})$ can be written as

$$n(\vec{r}) = n_0 + n_1(\vec{r}) \tag{3.1}$$

where $n_0 = \langle n(\vec{r}) \rangle \cong 1$ is the mean value of the index of refraction of air at atmospheric pressure and $n_1(\vec{r})$ is the random variation of $n_1(\vec{r})$ from its mean value, with a zero average $n_1(\vec{r}) = 0$.

The covariance function of $n(\vec{r})$ is given by

$$B_n(\vec{r}_1, \vec{r}_2) = B_n(\vec{r}_1, \vec{r}_1 + r) = \langle n_1(\vec{r}_1) n_1(\vec{r}_1 + \vec{r}) \rangle + n_0^2 \tag{3.2}$$

where $\vec{r}_1$ and $\vec{r}_2$ are two points in space, and $\vec{r} = \vec{r}_2 - \vec{r}_1$.

The structure function:

$$D_n(r) = \left\langle \left[n(r_1 + r) - n(r_1) \right]^2 \right\rangle = 2 \left[B_n(o) - B_n(r) \right] \tag{3.3}$$

The structure function simplifies to the following Kolmogorov-Obukhov two-thirds power law for isotropic and homogeneous turbulence:

$$D_n(r) = C_n^2 r^{2/3} \quad l_0 \ll r \ll L_0 \tag{3.4}$$

where l_0 is the inner scale size, L_0 is the outer scale size, and C_n^2 is the index of refraction structure parameter.

The wave number spectrum within the inertial range is [4] given by

$$\Phi_n(k) = 0.033 C_n^2 k^{-11/3} \quad \text{for } 1/L_0 \leq k \leq 1/l_0 \tag{3.5}$$

Some other forms of the wave number spectrum are also considered:
Tatarskii spectrum:

$$\Phi_n(k) = 0.033 C_n^2 k^{-11/3} \exp\left(-k^2 / k_m^2\right), \quad k_m = 5.92/l_0 \tag{3.6}$$

Von Kármán spectrum:

$$\Phi_n(k) = 0.033 C_n^2 \frac{\exp\left(-k^2 / k_m^2\right)}{\left(k^2 + k_0^2\right)^{11/6}}, \quad \text{where } k_0 = 1/L_0 \text{ or sometimes } k_0 = 2\pi/L_0.$$

Hill spectrum:

$$\Phi_n(k) = 0.033 C_n^2 \left[1 + 1.802\left(k/k_1\right) - 0.254\left(k/k_1\right)^{\frac{7}{6}} \frac{\exp\left(-k^2/k_L^2\right)}{\left(k^2 + k_0^2\right)^{11/6}} \right] \tag{3.7}$$

$$0 \leq k < \infty \quad \text{and} \quad k_1 = 3.3/l_0$$

3.3.1.2   FSO Communication Parameters

The optical field $U(r, L)$ of the received signal on the detector located at a distance L from the transmitter end is converted to electrical signal by the photodetector which is then processed by signal processor to recover the information. The coherence portion of the field is represented by the first moment $\langle U(r, L) \rangle$, where $\langle \ \rangle$ denotes an ensemble average. The mutual coherence function (MCF) of the wave is then defined by the second moment and is written as

$$\text{MCF} \equiv \Gamma_2(r_1, r_2, L) = \langle U(r_1, L) U^*(r_2, L) \rangle \tag{3.8}$$

where r_1 and r_2 are two points on the receiver plane and $U^*(r, L)$ is the complex conjugate of $U(r_1, L)$.

The fourth-order moment or the cross-coherence function of the field is given by

$$\Gamma_4\left(r_1,r_2,r_3,r_4, L\right)=\left\langle U\left(r_1,L\right)U^*\left(r_2,L\right)U\left(r_3,L\right)U^*\left(r_4,L\right)\right\rangle \tag{3.9}$$

From the second moment of irradiance and the mean irradiance, the scintillation index can be determined:

$$\left\langle I^2\left(r, L\right)\right\rangle=\Gamma_4\left(r,r,r,r, L\right) \quad \text{and} \quad \left\langle I\left(r, L\right)\right\rangle=\Gamma_2\left(r,r, L\right) \tag{3.10}$$

Scintillation index:

$$\sigma_I^2\left(r,L\right)=\frac{\left\langle I^2\left(r, L\right)\right\rangle}{\left\langle I\left(r, L\right)^2\right\rangle}-1 \tag{3.11}$$

Rytov variance is a parameter generally used to classify the various turbulence conditions:

$$\sigma_R^2 = 1.23 C_n^2 k^{7/6} L^{11/6}$$

$\sigma_R^2 < 1$ (weak fluctuations), $\sigma_R^2 \approx 1$ (moderate fluctuation), $\sigma_R^2 > 1$ (strong fluctuations), and $\sigma_R^2 \gg 1$, $\sigma_R^2 \to \infty$ (saturation regime).

A system approach for analyzing and evaluating FSO communication system performance metrics is shown in Fig. 3.4, which starts with a communication scenario and then provides communication inputs from transmitter, type of atmospheric/underwater channel, and mitigation techniques used to evaluate the FSO communication performance metrics, such as SNR, BER, and probability of fading [17, 18]. The general system approach is valid for analyzing FSO communication between ground and integrated terrestrial, satellite/aerial, or underwater terminals.

Some Communication System Performance Metrics Relevant to FSO Communication Systems

- Probability density functions of intensity fluctuations
- Received signal-to-noise ratio (SNR) and bit error rate
- Probability of fades

Free-space optical (FSO) communication links over several kilometers such as horizontal path to thousands of kilometers such as ground-to-satellite/space systems have increasingly attracted much attention in the past decade to develop technologies to provide high-bandwidth high-capacity wireless communication links. However, turbulence-induced effects due to atmospheric scintillation cause beam spreading and beam wander, which result in power loss and increased bit error rate

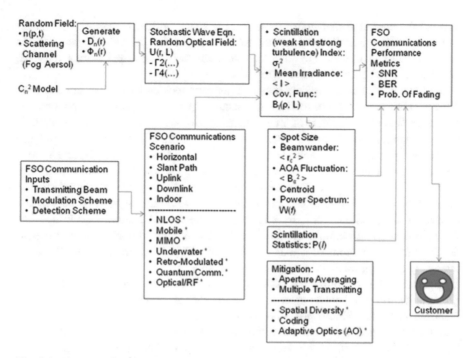

Fig. 3.4 Flowchart for FSO communications system performance evaluation. (*Reprinted with permission from Springer Nature and Copyright Clearance Center, 2021* [2])

due to signal fading in communication channels which ultimately degrade the link performance severely. In FSO communication technology applications, a Gaussian beam wave is the most practical way to focus, modify, and shape the easily available diode laser beam using lenses and other optical elements. The Gaussian beam is a beam of monochromatic electromagnetic radiation whose amplitude envelope in the transverse plane is given by a Gaussian function which is also true for Gaussian intensity (irradiance) profile. The fundamental (or TEM$_{00}$) transverse Gaussian mode describing the intended output of most lasers and the beam can be focused into the most concentrated spot: an important criterion for received power in a FSO communication system. It is therefore important to understand the basic optical propagation models of a Gaussian beam wave when transmitted and received by a communication system.

Uplink and Downlink Wave Models for FSO Communication System Performance

Applications of FSO communication systems include up-and-down slant-path links between space platforms and aircraft, ships, and other ground platforms, satellite-to-satellite cross-links, and also among mobile or stationary/fixed terminals for solving the so-called *last-mile* problem. Free-space optical wireless communications have also potential applications to develop the current 5G/6G networks. In fact, space/air terminals will also be integrated using slant-path links with the mobile and fixed 5G/6G system terminals.

To predict FSO communication system performance under a wide range of conditions requires underlying models that accurately capture system component performance under a wide range of atmospheric conditions, which include atmospheric optical channel model. For a space/aerial-based terminal, it is necessary to use uplink and downlink atmospheric channel models that accurately capture system and component performance between ground and space/aerial terminals. Optical wave models necessary to evaluate the FSO communication system performance for slant paths applicable to ground-to-satellite or satellite-to-ground links are discussed in an excellent book by Larry C. Andrews and Ronald L. Phillips [3]. Altitude-dependent channel models should also include the zenith-angle dependence in the calculations of signal-to-noise ratio (SNR) and bit error rate (BER) to evaluate FSO communication system performance.

Laser satellite communication systems use high-data-rate optical transceivers for communication channels connecting ground/airborne-to-space or space-to-ground/airborne data links. Since optical beams are affected by the atmospheric turbulence and scattering media, optical scintillations of the transmitted and received signals ultimately limit the communication capacity for high-data-rate FSO communications. For FSO communication system design point of view, it is therefore important to evaluate the signal fading below a prescribed threshold value using accurate optical models for predicting scintillation levels. For practical situations of both uplink and downlink laser satellite communication, optical scintillation for both weak and moderate-to-strong fluctuation theory and for zenith angles to 45°–60° and greater is necessary.

Basic Primary Optical Propagation Parameters for FSO Uplink and Downlink Communications

The basic primary parameters for both uplink and downlink ground/satellite communication are the following:

(a) Scintillation index (weak and weak-to-strong fluctuations)
(b) Spatial coherence radius
(c) RMS angle of arrival and beam wander
(d) Effective spot size
(e) RMS beam wander
(f) On-axis scintillation index (tracked and untracked)

FSO communication system designer requires to evaluate these parameters for a particular scenario and the FSO communication subsystems for selecting the optimum components for designing efficient laser/optical transceivers to operate under various channel links.

An extensive amount of research papers have been reported for the last few years. Most of the results discussed in this section can be found in some excellent books [1–3]. A summary of the theory of FSO communication signal propagation through atmospheric channel is discussed below.

Using the basic Gaussian beam wave with unit amplitude, the Gaussian beam parameters are characterized by Andrews and Phillips [3] and are repeated for reference to our slant-path propagation analysis:

Input-plane beam parameters:

$$\theta_0 = 1 - \frac{L}{F_0} \tag{3.12}$$

$$\Lambda_0 = \frac{2L_0}{kW_0^2}$$

where W_0 and F_0 are the beam radius and phase front radius of curvature at the transmitter output aperture.

The *output parameter* in the receiver plane at $z = L$:

$$\Theta = 1 + \frac{L}{F} = \frac{\theta_0}{\theta_0^2 + \Lambda_0^2}$$

$$\overline{\Theta} = 1 - \Theta \tag{3.13}$$

$$\Lambda = \frac{2L}{kW^2} = \frac{\Lambda_0}{\theta_0^2 + \Lambda_0^2}$$

where W and F are the beam radius and phase front radius at the receiver plane, respectively.

For FSO communication system, the signal received at the receiver determines the available power, which also determines the communication performance parameters such as SNR, BER, and fading statistics.

As mentioned earlier, Gaussian beams are normally used for practical purpose and availability because of the mature technology of diode lasers and the associated optics to design the communication system. The *free-space irradiance profile* of the Gaussian beam at the receiver plane:

$$I^0\left(r,L\right) = \left(W_0^2 / W^2\right) \exp\left(-2r^2 / W^2\right) \tag{3.14}$$

Mean Irradiance and Beam Spreading Due to Turbulence

Mean irradiance is one of the most important parameters which determine the power in the bucket (PIB) at the receiver to design a high-capacity communication system. By adjusting the laser power, transmitting, and receiving profiles by appropriately choosing the transceiver optics, the system performance can be greatly improved.

The mean irradiance is given by

$$\langle I(r, L)\rangle = \Gamma_2(r,r, L) = \frac{w_0^2}{w^2}\exp\left(-\frac{2r^2}{w^2}\right)\exp\left[2\sigma_r^2(r, L)-T\right] \qquad (3.15)$$

The mean irradiance can be evaluated from the second moment of irradiance, Eq. (3.10).

For a Kolmogorov spectrum

$$\Phi_n(k,h) = 0.033 C_n^2(h)k^{-11/3}$$

where $C_n^2(h)$ is the refractive index structure parameter and represents the strength of turbulence as a function of altitude h, which varies from the altitude of the transmitter h_0 to that of the receiver H. For the slant path, the propagation length is $L = (H - h_0)\sec(\zeta)$ with zenith angle ζ. Note that the turbulence is negligible above the tropopause layer with altitude $H_0 = 20$ km, i.e., $C_n^2(h > H_0) = 0$ for practical purposes.

For a free-space laser satcom system with $H > H_0$, H can be replaced by H_0 in r_0, the coherence parameter (see below), and, thus, r_0 mainly depends on k and ζ.

A number of profile models have been discussed for the last many years for the refractive index structure parameter $C_n^2(h)$; see for example Refs. [1, 9]. Mutual coherence function (MCF) discussed earlier is closely related to the wave structure function (WSF), which is dominated by the phase structure function.

Scintillation Index

According to the first-order Rytov theory, the scintillation index σ_I^2 can be expressed as the sum of the radial $\sigma_I^2(r,L)$ and longitudinal $\sigma_I^2(L)$ components. In the absence of optical turbulence, additional diffraction and refraction cause further broadening of the beam spot size as well as further focusing and defocusing.

The variance of intensity fluctuation of a satellite optical communication determines the SNR and therefore BER to determine the performance of satellite FSO communication system performance. When the satellite altitude H is much greater than 20 km, the irradiance fluctuation scintillation index $\sigma_I^2(r,L)$ can be simplified for satellite-to-ground path taking account of the zenith angle as follows [12]:

$$\sigma_I^2(r, L) = 2.25 k^{7/6}\sec^{\frac{11}{6}}(\zeta)\int_{h_0}^{H} C_n^2(h)(h-h_0)^{5/6}\,dh \qquad (3.16)$$

The aperture average factor can be included in the scintillation index expression [9, 12] when the receiver aperture is larger than the transverse irradiance correlation width. In addition, various adaptive optics techniques [1] can be applied to minimize the scintillation index to increase the received SNR and lower the BER in the performance of the satellite laser communication system performance.

As mentioned earlier in this chapter, one of the major degradations in the performance of the free-space optical (FSO) communication system is the intensity fluctuations caused by atmospheric turbulence. In addition, atmospheric turbulence also leads to the loss of spatial coherence of the propagating optical beam transmitted from a coherent source such as laser diode-emitting Gaussian beam wave. For a satellite/ground terminal, uplink or downlink scintillation degrades the communication performance parameters such as received signal-to-noise ratio (SNR), bit error rate (BER), and probability of fade statistics. This is particularly true for a long-range slant path like satellite-to-ground (downlink) and ground-to-satellite (uplink) situations where there is still no correct and exact optical turbulence profile because of (1) nonuniform turbulence along the slant path consisting of isotropic and non-isotropic turbulence and (2) the turbulence profile being dynamic and changing consistently with time because of constant change of wind velocity, temperature, and somewhat pressure along the line-of-sight (LOS) propagation link for the satellite changing positions. The nature of the problem remains the same not only for satellites but also for other space/aircraft like HAPs, UAVs, or aircraft moving at lower altitudes. Therefore, in principle, a real-time or almost real-time monitoring of the atmospheric scintillations along the propagation path would be needed in order to mitigate the atmospheric turbulence effects and thus optimize the FSO communication system parameters and performance. A recent paper [15] by the author and his colleagues discussed a new target-in-the-loop (TIL) atmospheric sensing concept for in situ remote measurements of major laser beam propagation characteristics and atmospheric turbulence parameters, which allows retrieval of key beam and atmospheric turbulence characteristics including scintillation index and refractive index structure parameter. The technique allows to measure intensity scintillation index $\sigma_I^2\left(I\left(x,y,z\right)\right)$ as a function of the moving satellite or space/aerial aircraft trajectory $I(x, y, z)$ and hence reconstruct from these measurements the corresponding refractive index profile $C_n^2\left[I\left(x,y,z\right)\right]$. A brief description of the method is described below, and interested readers are encouraged to read the paper [15] for full details.

TIL Laser Beam and Atmospheric Sensing

The concept target-in-the-loop (TIL) relies on integral relationship between complex amplitudes of the counter-propagating optical waves, and their values are preserved along the propagation path, for example along the slant path from ground to a spacecraft for a FSO communication system. For analysis purpose, the system configuration is composed of a single-mode fiber-based optical transceiver and a

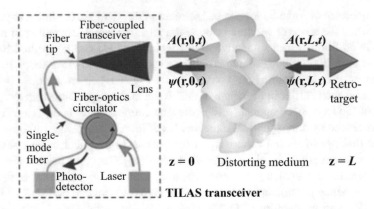

Fig. 3.5 An illustration of the sensing geometry with the atmospheric sensing transceiver at the pupil plane at $z = 0$ operating with a retro-target at the plane $z = L$. (*Reprinted with permission from The Optical Society of America, OSA, 2021* [15])

remotely located small-sized retroreflector, which can be easily designed to be placed at both ends of a ground/satellite FSO communication system. The interference metric is given by [15]

$$J_{int}(t) = \int A(r, z, t)\psi(r, z, t)d^2r$$
$$= \text{const}, \quad 0 \leq z \leq L \tag{3.17}$$

where $A(r, z, t)$ and $\psi(r, z, t)$ are the complex amplitudes of the outgoing and returned waves, respectively; $\mathbf{r} = \{x, y\}$ is a vector in the plane orthogonal to the optical axis z; and t is the time, the integration being performed over the whole transverse plane. Based on the optical reciprocity principle [16], and noting the conservation of the remote sensing invariant, the interference metric has identical value along the propagation path regardless of the propagation distance along the line of sight (LOS) between the transceiver aperture and the target. The concept is illustrated in Fig. 3.5, consisting of a transceiver telescope and an optical train based on a single-mode fiber and fiber elements used for delivering a collimating Gaussian beam generated at the fiber tip to the transceiver telescope pupil and delivering the target-return optical wave to the fiber-coupled photodetector.

The signal measured by the photodetector, $P(t)$, is related to the square of the interference metric and is given by

$$P(t) = \alpha \left| J_{int}(t) \right|^2 = \alpha \left| \int M_0(\mathbf{r})A(\mathbf{r}, 0, t)\psi(\mathbf{r}, 0, t)d^2\mathbf{r} \right|^2$$
$$= \alpha \left| \int M_T(\mathbf{r})A(\mathbf{r}, L, t)A(-\mathbf{r}, L, t)d^2\mathbf{r} \right|^2, \tag{3.18}$$

where $M_0(r)$ is the stepwise aperture function and $M_T(r)$ is the target aperture function. The signal measured by the transceiver can be written as

$$P_0(t) = \alpha \left| \int M_0(r)A(r, 0, t)\psi(r, 0, t)d^2r \right|^2$$

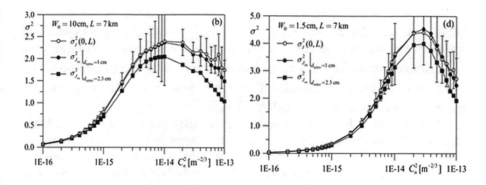

Fig. 3.6 Scintillation index σ_I^2 (white circles) and normalized variation of interference metric $\sigma_{J_{int}}^2$ (black circles and squares) for different turbulence strength C_n^2 values for collimated Gaussian beam of radius 10 and 1.5 cm and for propagating distance of 7 km. (*Reprinted with permission from The Optical Society of America, OSA, 2021* [15])

and the square modulus of the interference metric at the target plane as

$$P_T(t) = \alpha \left| \int M_T(r) A(r,L,t) A(-r,L,t) d^2r \right|^2 \tag{3.19}$$

which proves that $P_0(t) = P_T(t)$.

For the target located on the optical axis, the target aperture function can be described by $M_T(r) = \delta(r)$ which when substituted in the above equations gives

$$P_0(t) = \alpha \left| A^2(0, L, t) \right|^2 = \alpha I^2(0, L, t) \tag{3.20}$$

The instantaneous value of the on-axis signal intensity of the transmitted beam at the target is thus related to the square root of the measured signal, $P_0(t)$, in the following way: $\sqrt{P_0(t)} \sim I(0, L, t)$.

The above Eq. (3.20) provides a possibility of direct remote sensing of the laser beam intensity fluctuation variance, on-axis scintillation index, $\sigma_I^2(0,L)$. The normalized intensity variance can be written as

$$\sigma_I^2(0, L) \equiv \frac{\langle I^2(0, L, t) \rangle}{\langle I(0, L, t) \rangle^2} - 1 = \frac{\langle P_0(t) \rangle}{\langle \sqrt{P_0(t)} \rangle^2} - 1 \tag{3.21}$$

where $\langle \, \rangle$ is the ensemble statistical average. The intensity scintillation index can be approximated by $\sigma_I^2 \approx \sigma_{J_{int}}^2 = \langle P_0 \rangle / \langle \sqrt{P_0} \rangle^2 - 1$, where $\sigma_{J_{int}}^2$ denotes the variance of the interference metric, J_{int}, as defined earlier. Figure 3.6 shows the results of wave optics simulations that compare the values of the σ_I^2 calculated numerically at the target plane with the corresponding estimates $\sigma_{J_{int}}^2$ obtained by the signal $P_0(n)$ measurements and demonstrate that the normalized interference metric variance $\sigma_{J_{int}}^2$ obtained from the $P_0(n)$ signal measurements using the target-in-the-loop system provides accurate estimation of the "true" value of σ_I^2 for a variety of turbulence strength conditions denoted by C_n^2 values.

The technique is based on an optical system (transceiver) combining a laser/receiver pair with a retroreflector. Thus, a retroreflector or a small-diameter sphere with a set of retroreflectors conformal to its surface can be attached to a satellite, HAP, UAV, or airborne platform to remotely sense the scintillation index, which is required to develop a mitigating and compensating atmospheric turbulence along a long slant-path FSO communication system. The technique will thus allow to eventually improve FSO communication system performance.

Scintillation Index (Moderate-to-Strong Turbulence) for a Three-Layer Altitude-Dependent Atmospheric Turbulence

The turbulence in the troposphere and stratosphere obeying non-Kolmogorov statistics has been recently reported [9]. A more accurate three-layer atmospheric model was also developed [10]. The three-layer model consists of (1) boundary layer (up to about 1–2 km), (2) free troposphere (up to about 8–10 km), and (3) stratosphere above them whose turbulence power spectrum law exponents are 11/3 (Kolmogorov), 10/3 (non-isotropic), and 5 (non-isotropic), respectively, in the three layers. It is therefore necessary to investigate the irradiance and angle-of-arrival fluctuations of the Gaussian optical beam relevant to FSO communication systems to include both weak and moderate-to-strong turbulence encountered in typical slant-path satellite communication system. The turbulence spectrum of the three-layer turbulence model can be expressed as [11]

$$\Phi(\kappa, \alpha, h) = A(\alpha_i)\beta_i(h)\kappa^{-\alpha_i}, \quad i = 1,2,3, \tag{3.22}$$

where α_i is the power law exponent of the turbulence spectrum, κ is the spatial wave number, $\kappa = 2\pi/\lambda$, and h is the altitude. The constant $A(\alpha)$ is given by

$$A(\alpha) = \sin\left[(\alpha - 3)\pi/2\right] \cdot \Gamma(\alpha - 1)/4\pi^2, \quad 3 < \alpha < 5,$$

where $\Gamma(*)$ is the gamma function. The parameter $\beta(z)$ is the general refractive index structure constant (units of $m^{3-\alpha}$) and is given by [11]

$$\beta(h) = \left[A(11/3)/A(\alpha)\right]C_n^2\left(k/L\right)^{1/2(\alpha-11/3)} \tag{3.23}$$

Scintillation Index for Gaussian Beam for Three-Layer Model: Weak Turbulence

The scintillation index for the three-layer model can then be expressed as

$$\sigma_I^2 = \sum_{i=1}^{3} \sigma_{R\_i}^2 = \sum_{i=1}^{3} 4\pi^2 A(\alpha_i)\Gamma\left(1 - \frac{\alpha_i}{2}\right) k^{3-\frac{\alpha_i}{2}} \sec(\zeta) L^{\frac{\alpha_i}{2}-1} \mu_{1\_i}(\Lambda,\Theta,\alpha_i), (3.24)$$

where the contributions of the scintillation index for the boundary layer, free troposphere, and stratosphere are

$$\sigma_{R\_1}^2, \sigma_{R\_2}^2, \text{ and } \sigma_{R\_3}^2,$$

respectively.

The parameters $\mu_{1\_i}(\Lambda, \Theta, \alpha_i)$ in the above equation are

$$\mu_{1\_i}(\Lambda,\Theta,\alpha) = \text{Re}\left(\int_{h_{0\_i}}^{H_i} \beta_i(h)\left\{ \Lambda^{(\alpha_i-2)/2}\xi^{(\alpha_i-2)} \right. \right.$$
$$\left. \left. -\xi^{(\alpha_i-2)/2}\left[\Lambda\xi + i(1 - \bar{\Theta}\xi) \right]^{(\alpha_i-2)/2} \right\} dh \right) \quad i = 1,2,3,$$
(3.25)

where $i - 1$, 2, and 3 represent three boundary layers with H_1 being the top of the boundary layer, H_2 the top altitude of the second boundary layer, and H_3 the top of free troposphere which is equal to H. The results [11] show that the scintillation index for the three-layer model is greater than the conventional Kolmogorov model and the scintillation indices of different Gaussian beams in the *downlink* path are almost the same but are very different from the *uplink* path. The scintillation index results for moderate-to-strong turbulence are reported [11], and the readers are encouraged to study the detailed derivations and the final results for more details to apply to design FSO communication system such as LEO satellite using Gaussian laser beam as well as for different space/aerial-based scenarios involving slant paths. It is also important to develop better design for an optimal Gaussian beam to mitigate the intensity scintillation specially for an uplink path.

Spatial Coherence Radius

A Gaussian beam wave for a *downlink* path is given by the equation [12]

$$D(\rho,L) = 2.91k^2\rho^{5/3}\mu_0\sec(\zeta)$$

where

$$\mu_0 = \int_{h_0}^{H} C_n^2(h)dh$$
(3.26)

where h_0 is the height aboveground level of the uplink transmitter and/or downlink receiver, $H = h_0 + L \cos(\zeta)$ is the satellite altitude, and ζ is the zenith angle. The other quantity of interest to the satellite-to-ground propagation path is the *spatial coherence radius*, which is the distance ρ, where $D(\rho, L) = 2$ which after substitution in the above equation is given by

$$\rho_0 = \left(\frac{\cos \zeta}{1.45 \mu_0 k^2} \right)^{3/5}$$

Another parameter r_0 called the Fried's parameter is defined by the atmospheric coherence width $r_0 = 2.1\rho_0$:

$$r_0 = \left[0.42 \sec(\zeta) k^2 \int_{h_0}^{H} C_n^2(h) dh \right]^{-3/5}$$

For an uplink propagation path, the spatial coherence radius is given by [12]

$$\rho_0 = \left(\cos \zeta \left| 1.46 k^2 \left(\mu_{1u} + 0.622 \mu_{2u} \Lambda^{11/6} \right)^{3/5} \right. \right.$$ (3.27)

where Λ, the Gaussian beam parameter, was defined earlier and

$$\mu_{1u} = \int_{h_0}^{H} C_n^2(h) \left(\Theta + \overline{\Theta} \left(\frac{h - h_0}{H - h_0} \right) \right)^{5/3} dh$$ (3.28)

and

$$\mu_{2u} = \int_{h_0}^{H} C_n^2(h) \left(1 - \left(\frac{h - h_0}{H - h_0} \right) \right)^{5/3} dh$$ (3.29)

RMS Angle of Arrival and Beam Wander

For FSO communication system, the wave front of the laser beam at the receiver will be distorted due to atmospheric turbulence. When the optical signal is focused by the receiver aperture, wave-front distortions will cause to spot motion or image dancing (jitter) at the focal plane of the receiver. Angle-of-arrival fluctuations can be described by phase structure function. This is called angle-of-arrival fluctuations, which can be compensated by the adaptive optics (AO) or fast steering mirror. *Angle-of-arrival fluctuation* is an important parameter for both uplink and downlink paths. For a receiver aperture of diameter $2W_G$, the variance of the angle-of-arrival fluctuations is given by [3]

$$\text{Downlink: } \langle \beta_a^2 \rangle \cong 2.91 \mu_0 \sec(\zeta)(2W_G)^{-1/3} \tag{3.30}$$

$$\text{Uplink: } \langle \beta_a^2 \rangle \cong 2.91 \left(\mu_{1u} + 0.62 \mu_{2u} \Lambda^{11/6} \right) \sec(\zeta)(2W_G)^{-1/3} \tag{3.31}$$

where the Gaussian beam parameters are defined earlier. For downlink path, the variance of the angle-of-arrival fluctuations is independent of the wavelength and is typically on the order of several μrad, whereas for an uplink it is ~<1 μrad. FSO system using line-of-sight (LOS) communication therefore requires proper system design parameters such as transmitter beam radius, beam divergence, transmitter power, operating wavelength, and receiver aperture size for selecting the FOV for improving communication system performance.

For FSO laser communication system design optimization of transmitter beam radius, beam wander caused by atmospheric turbulence-induced scintillation is very important. Essentially, beam wander is the effect causing "wandering" of the laser spot leading to a random distribution of the radial distance from the center of the spot. Beam wander therefore affects the performance of the *laser satcom* system.

When the size of turbulence eddies is larger than the transmitting beam size, random deflection of the beam from the on-axis occurs when propagating through the atmospheric turbulence. The phenomenon is termed as beam wander and will lead to link failure. This is more prominent for uplink and will result in beam displacement by several hundred meters. The RMS beam wander displacement is given by [1]

$$\sigma_{BW}^2 \approx 0.54 (H - h_0)^2 \sec^2(\zeta) \left(\frac{\lambda}{2W_0} \right)^2 \left(\frac{2W_0}{r_0} \right)^{5/3} \tag{3.32}$$

where h_0 is now the altitude of the transmitter.

While beam wander and angle of arrival at the receiver plane are the main contributions for signal degradation for the uplink, beam spreading, scintillation, and loss of spatial coherence are the dominant contributions for the downlink.

Recently, beam wander on uplink ground-to-satellite laser communication has been reported [12] for a collimated beam in a weak turbulence condition. In their research, the scintillation index σ_I^2 was expressed as the sum of the radial $\sigma_{I,r}^2(r,L)$ and longitudinal $\sigma_{I,l}^2(r,L)$ components. After simplifying the expressions of the parameters derived in Andrews and Phillips [3], the beam wander is given by Guo et al. [12] for larger distance $(H \gg H_0)$,

$$\begin{aligned}
\sigma_{BW}^2 &= 5.95 (H - h_0)^2 \sec^2(\zeta) \left(\frac{2W_0}{r_0} \right)^{5/3} \left(\frac{\sigma_{pe}}{WL} \right)^2 \\
&= 79.9 \left(\frac{W_0}{r_0} \right)^{10/3} \left[1 - \left(\frac{4\pi^2 W_0^2 / r_0^2}{1 + 4\pi^2 W_0^2 / r_0^2} \right)^{1/6} \right],
\end{aligned} \tag{3.33}$$

From Eq. (3.33), it is clear that the beam wander increases with W_0, and therefore an optimum size of W_0 will improve the FSO communication system performance.

Mean irradiance at the pupil plane of the receiver can be approximated by a Gaussian spatial profile and is expressed as [12]

$$\langle I(r, L)\rangle \cong \frac{w_0^2}{w_{LT}^2}\exp\left(-\frac{2r^2}{w_{LT}^2}\right)\left[w/m^2\right]$$

where w_{LT} is the effective or long-term spot size of the Gaussian beam in the presence of atmospheric optical turbulence and is given by

$$W_{LT} = W\left[1 + 4.35\mu_{2d}\Lambda^{\frac{5}{6}}k^{\frac{7}{6}}(H-h_0)^{\frac{5}{6}}\sec^{\frac{11}{6}}(\zeta)\right]^{1/2} \tag{3.34}$$

The downlink parameter μ_{2d} can be written as

$$\mu_{2d} = \int_{h_0}^{H}C_n^2(h)\left(\frac{h-h_0}{H-h_0}\right)^{5/3}dh$$

Uplink Scintillation Index and Beam Wander

In an optical communication link between an optical ground station and a LEO or GEO satellite, the *uplink* is the main problem which arises due to scintillation and beam wandering. The robustness of the communication performance will depend on the accurate channel models and the simultaneous effects of the scintillation and beam wander. Larger atmospheric turbulence eddies crossed by the entire optical beam cause some degrees of refraction, which then deviates the original beam from the boresight of the original path known as *beam wander* and will lead to link failure. In an FSO communication system, beam wandering thus has important impact on the received time-varying power and the fading statistics. For *uplink*, the beam wander can have significant effects of the order of a several hundred meters which can limit the FSO communication system performance and therefore most efficient designs of optical/laser transceivers will be required. Beam wander effect is negligible for *downlink* because the turbulence eddies much smaller than the beam size do not shift the beam's centroid significantly when reaching the FSO receiver plane at the optical ground station. However, effects of angle-of-arrival fluctuations cause the wave-front tilt at the receiver, which requires some mitigation technique to increase the SNR and decrease the BER of the system.

The RMS beam wander displacement is given by [13]

$$\sigma_{BW}^2 \approx 0.54(H-h_0)^2\sec^2(\zeta)\left(\frac{\lambda}{2W_0}\right)^2\left(\frac{2W_0}{r_0}\right)^{5/3} \tag{3.35}$$

where h_0 is now the altitude of the transmitter.

The long-term spot size for a ground-to-satellite optical path is useful to design the receiver aperture and detection methods and can be written as

$$W_{LT} = \sqrt{1 + \left(\frac{D_0}{r_0}\right)^{5/3}} \qquad \text{for } 0 \le \frac{D_0}{r_0} < 1$$

$$= W\left[1 + \left(\frac{D_0}{r_0}\right)^{\frac{5}{3}}\right]^{\frac{3}{5}} \qquad \text{for } 0 \le \frac{D_0}{r_0} < \infty$$

(3.36)

where r_0 is the atmospheric coherence width and $D_0 = 8w_0^2$.

For an uplink path, much more beam spreading occurs close to the transmitter at the ground location and therefore requires proper design so that the maximum intensity of the upward beam reaches the receiver located at the satellite.

For an uplink scenario, beam wander which is the angular deviation of the beam from the boresight is an important factor to be taken into account. Beam wander can cause time-varying power fades, which at times can be large enough to lose the Internet connectivity. The transmitting beam type and size such as of a collimator beam from ground to satellite can be selected and designed by calculating the RMS angular beam wander. The uplink scintillation index, the pointing error variance, and covariance function of irradiance for an uplink satellite laser communication are described elsewhere [12].

Angular anisoplanatism plays an important role in FSO laser satcom when a moving uncooperative satellite target is tracked by one beam and the other intercepting beam to the target satellite as shown in Fig. 3.7 for a point-ahead propagating geometry. The transit time from the satellite to the ground and back again has to be adjusted using an accurate link maintenance control and tracking algorithm. If the satellite is moving with a speed of V, the point-ahead angle is $\theta_P \cong 2V_T/c$, where c is the velocity of light. Typically, θ_P is on the order of 50 μrad and $\theta_P \gg \theta_0$ (isoplanatic angle). Accurate ATP is absolutely required in order to enable the flow of FSO communication data between satellites and ground stations. This will specially be true when FSO laser satcom involves a *constellation* of satellites which will be discussed in detail in a later chapter of this book. Acquisition, tracking, and pointing will also be required to design FSO laser satcom to establish communications between satellites such as between GEO and LEO, or even between LEO and small satellites belonging to a constellation network. To facilitate the ATP process, sometimes a beacon is placed in a satellite, and the beacon and communication beams still have to go through the atmosphere which will need to be adjusted for turbulence and scattering.

Free-space optical (FSO) communication links between space-based terminals and ground stations are becoming an important element of space/ground network infrastructure. Obviously, this will allow a major reduction in the number of Earth

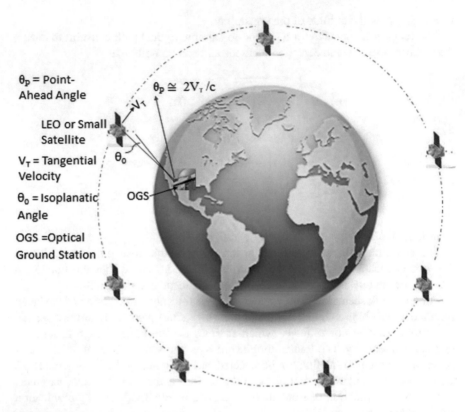

θ_P = Point-
Ahead Angle

$\theta_P \cong 2V_T/c$

LEO or Small
Satellite

V_T = Tangential
Velocity

θ_0 = Isoplanatic
Angle

OGS

OGS = Optical
Ground Station

Fig. 3.7 Point-ahead angle for FSO laser satcom showing the isoplanatic angle

stations needed to service the system. Using low Earth orbit (LEO) and geosynchronous orbits (GEO), there is potential future for providing satellite Internet in the Sky network at very high speed and increased communication capacity. As pointed out earlier, high data rate and small antenna size with a very narrow beam divergence offered by laser communications have a number of advantages for system design. Optical space communications using LEO and GEO satellites can cover a wide area network by providing cross-link between LEO and LEO or GEO and GEO with further capability to beam data to ground stations with links such as LEO-ground and GEO-ground. An illustration of various optical satellite communication (OSC) links where satellites and ground stations are connected by a one-way or a duplex laser link is shown in Fig. 3.8. Satellite 1 and Satellite 2 in the illustration can represent a geostationary, low Earth orbit or any other type of satellites such as a small, cube, or microsatellite, and the figure shows the concepts of optical satcom to establish the connectivity of the link. The possible links show how to establish connectivity in some cases where clouds can be an obstacle. It also shows the possibility of transmitting data from a particular ground location to another ground station located even in other part of the Earth via a laser link to satellites acting as *data relays*.

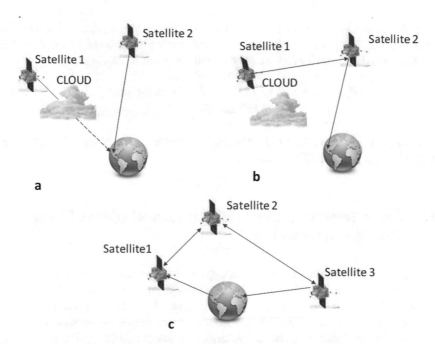

Fig. 3.8 Various link configurations used in optical satellite communication (OSC) showing point-to-point and cross-link systems. (**a**) To avoid cloud obstruction for Satellite 1 to connect to a ground station, Satellite 2 connects to the ground station. (**b**) As another way to avoid cloud obstruction, Satellite 1 relays the message to Satellite 2, which then sends back to ground station. (**c**) To send the communication message from a ground location to another part of the Earth ground location, laser links can be established via satellites acting as data relays

Laser-Based Free-Space Optical Links

Optical links in space can take advantage of an incredible number of high-technology achievements, which is obviously evidenced in the developments of optical terrestrial networks. Recent developments for FSO communications applications include transmitter (source), receiver (detector), optical telescopes using adaptive optics (AO), and MEMS-based agile beam steering, and also advanced software-based communications signal processing. For space-based FSO communication systems on board a satellite or other space platforms, size, weight, power, and of course cost are the main issues. For data rates of 10s of Gbit/s or more, it is helpful that the coherent beam transmitting from the source can be focused to spot sizes of the order of wavelength of the laser light which approach the diffraction limit. A narrow beam is particularly important in large LEO or small satellite constellation optical communication links since the potential interference to or from the adjacent satellites is greatly reduced. Otherwise, interference would cause and will contribute to the received signal noise affecting the available SNR and BER communication performance parameters. Footprint sizes for GEO and LEO orbits are already discussed in Chap. 1. For example, if the laser beam is focused from a GEO satellite directly along the equator, it will be a circle with a diameter (footprint) $\cong$784 m (GEO laser footprint). For LEO to ground, it is $\cong$34.72 m (LEO laser footprint) using a typical lasercom wavelength of $\lambda = 1.55$ μm and a transmitting diameter of 10 cm. For inter-satellite link (ISL), the footprint can be calculated from the equation

$$\theta_{GEO-LEO} = 2.28 * \lambda / D, \text{ and the footprint for}$$
$$GEO-LEO \text{ satellites} = R_{GEO-LEO} * \theta_{GEO-LEO}$$

where $R_{GEO-LEO}$ is the GEO-LEO inter-satellite distance, which depends on the altitude of the particular LEO selected.

3.4   Laser Satellite Communication Technologies and Link Design Fundamentals

This section will describe the basic concept of the functions of an optical terminal on board a spacecraft. In particular, the requirements of data transmitters and receivers as well as optical antennas and pointing, acquisition, and tracking mechanism will be briefly discussed. Basic space laser communications with application scenarios and brief descriptions of the systems, technologies, and applications are discussed. Figure 3.9 depicts a general architecture of integrated satellite/aerial network scenario, which includes both GEO and LEO satellites.

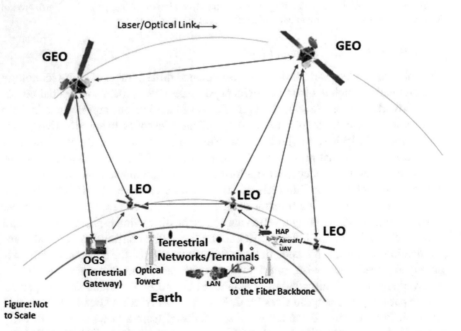

Fig. 3.9 Various laser communication links in space and in atmosphere. (Figure: Not to scale)

3.4.1 General Space Laser Communication System Technologies

For free-space optical (FSO) communications, generally a point-to-point data transfer between two spacecraft is required. A line of sight (LOS) is necessary for laser communication, and the communication distance between the two spacecrafts depends on the scenario. For example, between GEO and LEO orbit satellites, distance may be typically from a few hundred km to 70,000 km for near-Earth applications. Laser communications have the potential of providing data rates in the ranges of 10s of Gbit/s. A general architecture of satellite and high-altitude platforms (HAPs) is depicted in Fig. 3.10, showing concepts of downlink and uplink from satellite to the optical ground station (OGS) and direct optical link between HAPs.

3.4.1.1 Laser Transceiver Design

Laser transceivers need to be designed for the terminals in order to deliver bidirectional communication links. Usually, the transmitter and the receiver in a transceiver system share the same optical antenna, which is specially designed compound optical lenses. The two main design parameters obviously are transmitting beam angle (which should be very small in the range of a few microradians or so) and field of view (FOV) of the receiver so that photons from the receiving signal can be optimally focused to the photodetector for the best received signal-to-noise ratio (SNR). For inter-satellite links (ISLs), the beam steering or pointing capability with sub-microradian angular resolution and possibly with an angular coverage exceeding a hemisphere is critical in the space laser

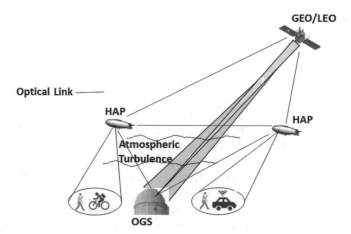

Fig. 3.10 A general architecture of satellite and high-altitude platforms (HAPs) showing concepts of downlink and uplink from satellite to the optical ground station (OGS) and direct optical link between HAPs

communication system design. A laser terminal performs three basic tasks, namely the data transmission aspects; pointing, acquiring, and tracking (PAT) of very narrow laser beams; and accurate interfacing with spacecraft platform with proper space-qualifiable opto-mechanical structures. The sequences for the operation of a transceiver system in general are the following:

- Transmitted data modulates the laser source and passes an optical duplexer and a fine pointing assembly before it enters an optical telescope (for both transmitting and receiving optical signals) where a course pointing assembly provides the steering of the optical antenna.
- Received signal also retracks back through the same antenna and the fine pointing assembly to the receive part of the terminal.
- A part of the received beam goes to a data detector, while the other part is used for controlling the fine and course pointing. A point-ahead system is also included in alignment between the transmission and reception.

One of the important considerations for sustaining uninterrupted communication using one common optical antenna for both transmit and receive beams is the proper design of a duplexing element able to separate the outgoing and incoming beams. For example, for a laser transmitting a power of 500 mW and a receive power of 5 nW, a 90 dB or more degree of isolation is necessary so that the cross talk can be negligible [14]. Other considerations include the Doppler effect due to the relative motions of both terminals, and it is given by $\Delta f = \Delta_D/\lambda$, where Δ_D is the relative velocity along the line of sight, and λ is the carrier frequency of the optical wave. For example, for LEO-GEO optical link at an operating wavelength of $\lambda = 1.06\ \mu m$, the Doppler shift can be around ±7.5 GHz, which is negligible for direct detection (DD) system, but for a coherent (heterodyne) detection scheme, the large Doppler frequency shift must be compensated in the detection process. Finally, adjustment due to a point-ahead angle because of the relative angular velocity of the two terminals will be required for both GEO-GEO and LEO-GEO optical links. The concept of point-ahead angle is discussed in the section.

3.4.1.2 Laser Transmitter System

The main parameters of an optical source are wavelength, output power, transverse mode, spectral characteristics such as linewidth, polarization, integration capability with transmitting optical telescope for generating desired beam shapes and divergent angles, and high-speed modulation capability, just to name a few. Recently, high-power laser diodes (>10 W) have been available at wavelengths from the near infrared through roughly the 2000 nm region. Required optical power is determined by the link range between two different orbital satellites, or inter-satellite link distance of the satellites at same orbit and the desired data rate. Advanced modulation scheme can be applied to improve the data rates and communication capacity of the communication system.

3.4.1.3   Optical Receiver

Receiver sensitivity is an important consideration for space laser links with satellites. The receiver sensitivity is characterized by the parameter of detector sensitivity expressed in minimum number of photons/bit. This is discussed in detail in a later section. Sometimes, receiver sensitivity can be characterized as the minimum number of input photons per bit to achieve a bit error rate (BER) probability of 10^{-6}. A direct detection receiver or a coherent detection receiver could require different sensitive detectors. Optical receivers employing avalanche photodetector (APD) with optical preamplifier are often used for a few Gbit/s data transfer. In order to reach toward a quantum limit, the designer has to make the effects of other noise contributions negligible.

3.4.1.4   Pointing, Acquisition, and Tracking (PAT)

In order to establish space optical link for interorbital or inter-satellite links, accurate pointing and acquisition are absolutely necessary. The subject of pointing, acquisition, and tracking (PAT) is described in detail in some other chapters of this book. Some recent advances in pointing and tracking with no moving parts and use of fine pointing of laser beams by using laser arrays for applications to satellites are very attractive for these types of future applications.

Laser satellite networks using all-optical technologies can also serve mobile data transfer and interconnectivity at very high data rates.

3.4.1.5   Laser Satellite Communication Channels: Primary
            Atmospheric Effects

The atmosphere extends to approximately 700 km above the Earth's surface with the most concentration of particulates, and molecules in the first 20 km above the surface. When the electromagnetic wave passes through the atmosphere, it suffers power losses and wave-front distortions due to the interactions of the electromagnetic wave with the particles. As a result, the received signal of the laser satellite communication system is severely affected specially at optical wavelengths in the range of 0.6–1.55 µm. The effects of atmospheric turbulence on the laser satellite communication discussed here are classified as *clear-air turbulence* (CAT) and generally will be considered for the *free atmosphere*, which extends roughly to 40 km and can consist of troposphere, tropopause region, and stratosphere. The region very close to the ground is called the atmospheric boundary layer (ABL) extending to roughly 1–2 km above the Earth's surface. Atmospheric dynamics in the ABL is resulted by the interaction and heat exchange with the Earth's surface. Atmospheric turbulence profile in the *atmospheric boundary layer* is not well understood because it depends on a number of uncontrollable factors near the ground. That is why sometimes an optical terminal/optical ground station (OGS) is designed to be located at some height above the ground.

Some of the typical types of laser communication channels and the associated primary atmospheric effects are summarized below:

- *Satellite/ground (downlink)*: The laser communications in this scenario, which is *downlink* propagation, are affected mostly due to the atmospheric turbulence near the ground, and this is mostly due to the effects of *scintillations* and *angle-of-arrival fluctuations*.
- *Ground/satellite (uplink)*: A transmitted laser beam near the transmitter is mostly affected. The dominant effects are *scintillations* and *beam wander* (related to *beam pointing*).
- *High-altitude platform/aircraft/UAV/balloon*: The atmospheric effects depend mostly on the heights of the terminals and the turbulence profiles between ground and terminals. Aircraft *boundary layer* might need to have special attention in transmitting or receiving laser beams.

3.4.2 Laser Satellite Communication System Performance Parameters

Some of the parameters necessary to define the performance of the laser satellite communication systems are:

- Received signal-to-noise ratio (SNR)
- Bit error rate (BER)

There are basically four scenarios for evaluating laser satellite communication system as mentioned earlier:

- *Satellite/ground (downlink)*
- *Ground-satellite (uplink)*
- *High-altitude platform/aircraft/UAV/balloon*
- *Inter-satellite (between two satellites either in the same orbit or in different orbits)*

An accurate design of a laser satellite communication system involves various scenarios of the optical links and calculations of their power link budgets and limitations. The link budget is a calculation of all the gains and losses between the transmitter and the receiver in the presence of atmospheric communication channel. The subject of link budget computation with various components is discussed [1, 2].

3.4.2.1 Received Signal-to-Noise Ratio (SNR)

For free-space laser communication systems, the ultimate limit to the detectability of a weak signal is set by noise where the shot noise, background noise, and thermal noise contribute to the total at the receiver. For a lasercom system for digital communication, the goal is to transmit the maximum number of bits per second over the maximum possible range with the fewest errors. Message data electrical signals are converted to optical signals via a modulator. At the receiver, the received coded signal is decoded using a decision based on threshold, which thus determines the best performance in decoding the correct signal with the lowest probability of making a bit decision error to ultimately obtain bit error rate (BER). For laser satellite communication, the received signal exhibits additional power losses and random irradiance fluctuations. The mean SNR at the output of the detector in the presence of atmospheric turbulence can be written as [1]

$$\langle SNR \rangle = \frac{SNR_0}{\sqrt{\frac{P_{S0}}{\langle P_S \rangle} + \sigma_I^2(D)SNR_0^2}} \tag{3.37}$$

where SNR_0 is the signal-to-noise ratio in the absence of turbulence defined by

$$SNR_0 = \frac{i_S}{\sigma_N} = \sqrt{\frac{\eta P_S}{2h\nu B}} \tag{3.38}$$

i_S = signal current, σ_N = root mean square (RMS) noise current, P_S = signal power in watts, η = the quantum efficiency of the detector, h = Planck's constant ($h = 6.63 \times 10^{-34}$ J s), and B = bandwidth of the (detector) filter. For shot noise-limited operation, the background noise and the thermal noise can be ignored.

3.4.2.2 Relationship Between Signal-to-Noise Ratio (SNR) and Bit Error Rate (BER)

In the presence of atmospheric turbulence, the SNR is a fluctuating term and therefore the average value of SNR is to be taken. The mean SNR can be written as [1]

$$\langle SNR \rangle = \frac{SNR_0}{\sqrt{\frac{P_{S0}}{\langle P_S \rangle} + \sigma_I^2(D)SNR_0^2}} \tag{3.39}$$

where SNR_0 = the SNR in the absence of turbulence defined earlier, P_{S0} = the signal power received in the absence of atmospheric effects, $\langle P_S \rangle$ = the mean input signal power, and $\sigma_I^2(D)$ = aperture-averaged scintillation index on the photodetector of diameter D.

The bit error rate (BER) is the probability of incorrect bit identification by the decision circuit at the receiver. This is an important communication system performance parameter and is usually specified by the system requirement. A typical value of BER is in the range of 10^{-6} to 10^{-9}. Assuming the Gaussian distribution for both noise and signal plus noise, the probability of error in the absence of turbulence is given by [1]

$$\mathrm{BER}_0 = \frac{1}{2}\left(\frac{\mathrm{SNR}_0}{2\sqrt{2}}\right) \tag{3.40}$$

In the presence of optical turbulence, the probability of error is considered a conditional probability, averaged over the probability density function (PDF) of the random signal fluctuations to determine the unconditional mean BER. This is given by $\langle\mathrm{BER}\rangle$ as follows [19]:

$$\langle\mathrm{BER}\rangle = \frac{1}{2}\int_0^\infty p_\mathrm{I}(s)\,\mathrm{erfc}\left(\frac{\langle\mathrm{SNR}\rangle s}{2\sqrt{2}\langle i_\mathrm{s}\rangle}\right)ds \tag{3.41}$$

where $p_\mathrm{I}(s)$ = probability density function (PDF) of irradiance, i_S = the instantaneous signal current whose mean value is given by $\langle i_\mathrm{S}\rangle = \eta e\langle P_\mathrm{S}\rangle/h\nu$, $\langle P_\mathrm{S}\rangle$ = the mean signal power as defined earlier, and $\langle\mathrm{SNR}\rangle$ = the mean SNR in the presence of turbulence also as defined earlier. Generally, a gamma-gamma probability distribution is used for calculations. Note that the average bit-error-rate computation in the presence of atmospheric turbulence involves the calculation of aperture-averaged scintillation parameter, $\sigma_\mathrm{I}^2(D)$, as shown in the earlier equation for mean signal-to-noise ratio $\langle\mathrm{SNR}\rangle$.

3.4.2.3   Downlink Channel

The irradiance flux variance in the focal plane of a receiver is given by [3]

$$\sigma_\mathrm{I}^2(D) = 8.70 k^{\frac{7}{6}}(H-h_0)^{\frac{5}{6}}\sec^{\frac{11}{6}}(\zeta)$$
$$\times \mathrm{Re}\int_{h_0}^H C_n^2(h)\left[\left(\frac{kD^2}{16L}+i\frac{h-h_0}{H-h_0}\right)^{5/6}-\left(\frac{kD^2}{16L}\right)^{5/6}\right]dh \tag{3.42}$$

where L = distance from the transmitter to the receiver, ζ = zenith angle, $H = h_0 + L\cos(\zeta)$, and h_0 = height aboveground of the *uplink* transmitter and/or *downlink* receiver. The steps involved in calculating the average bit error rate are the following:

- Select the accurate probability density function (PDF) characterizing the irradiance fluctuation statistics, $p_\mathrm{I}(s)$.

- Calculate the mean signal-to-noise ratio $\langle SNR \rangle$, which is the quantity defined in the presence of turbulence: The power ratio $\dfrac{P_{S0}}{\langle P_S \rangle} \cong 1 + 1.63 \sigma_R^{\frac{12}{5}} \Lambda_1$ where σ_R^2 is the Rytov variance, the Gaussian beam parameter $\Lambda_1 = \dfrac{2L_1}{kW_1^2}$ and the parameter W_1 denote the beam radius in the front plane of the lens for the propagation of Gaussian beam, and L_1 is the distance from the laser for originating a Gaussian beam [3]. This power ratio is basically the *Strehl ratio*, which is an important parameter in characterizing an imaging system.
- Finally, compute the average BER from the above equation using the average value of the received signal current $\langle i_S \rangle$, where $\langle P_S \rangle$ is the received signal power evaluated from the geometry of the free-space laser communication system.

3.4.2.4 Uplink Channel

For uplink channel, the more practical parameters in accurately designing a FSO communication system are the effective spot size at the receive plane, RMS beam wander radius, and on-axis scintillation index (tracked and untracked). The detailed analysis and descriptions can be found in [3].

3.4.3 HAP-to-HAP Optical Link: An Example

High-altitude platforms have lower altitudes approximately around 20 km, thereby with a short roundtrip delay (low latency), and have stationary positions for a while over a particular area. The atmosphere is very thin at that altitude so that the refractive index structure parameter (the so-called C_n^2) is much smaller than at the ground level. This makes the HAPs appropriate for establishing cross-links. In order to calculate HAP link budget and associated communication system performance parameters like signal-to-noise ratio (SNR) and bit error rate (BER), slant paths between HAPs and optical ground station (OGS) need to be considered.

Recently, researchers presented results using their developed software to design HAP-to-HAP optical link [19] where two HAPs at an altitude of 20 km each and separated by 300 and 600 km with intensity modulation direct detection (IM/DD) scheme are considered. The most important atmospheric turbulence effects at this altitude and between two HAPs are scintillation and beam spreading. Their atmospheric turbulence channel models for refractive index turbulence strength (C_n^2) as a function of height (h) applicable to the HAPs are the following [19]:

- Best-case turbulence: $C_n^2 = 10^{-15.7 - (h/7350\text{ m})}$ for $h \geq 13$ km

- Worst-case turbulence: $C_n^2 = 10^{-14.5 - (h/7630\text{ m})}$ for $h \geq 13$ km

The mean bit error probability is given by [19]

$$\langle p_B \rangle = \int_0^\infty Q\left(\frac{\langle P_{rcvd} \rangle / P_{char} \cdot y}{1 + \sqrt{1 + \xi \cdot \langle P_{rcvd} \rangle / P_{char} \cdot y}} \right) \cdot f_y(y) dy$$

$$Q(x) = 0.5 \cdot \operatorname{erfc}\left(x / \sqrt{2} \right)$$

(3.43)

where $\langle P_{rcvd} \rangle$ = mean received power, $f_y(y)$ = probability density function (PDF) of the optical received power fluctuations normalized to the mean value $\langle P_{rcvd} \rangle$, and erfc = the complementary error function. The aperture average factor can be included in the probability of error equation. The authors reported the overall link margin of approximately 2.6 dB, the mean error probability of 2.64×10^{-8} for HAP-HAP distance of 300 km, and 3.76×10^{-8} for a distance of 600 km. The theoretical analysis can also be applied to other platforms at lower altitudes such as CubeSat or nanosatellite systems.

3.5 Summary

Some of the fundamentals of laser satellite communications such as systems and technologies necessary to design them are presented in this chapter. Various types of GEO and LEO orbits for various optical links are explained, which are necessary to establish all-optical global communications and connectivity. Essential results of optical propagation theory relevant to laser satellite communication are analyzed with emphasis on uplink and downlink wave models. A new technology concept for remote sensing of scintillation index is described. Laser communication technologies and link design fundamentals are discussed. The relation between average signal-to-noise ratio (SNR) and average bit error rate (BER) necessary for establishing laser satellite links is explained with an example of high-altitude platform (HAP)-to-HAP optical link. This chapter provides a background and a solid foundation to move on to the understanding of laser satellite constellation in the next few chapters.

References

1. Arun K. Majumdar, *Optical Wireless Communications for Broadband Global Internet Connectivity: Fundamentals and Potential Applications*, Amsterdam, Netherland Elsevier (2019).
2. Arun K. Majumdar, Chapter 2, *Advanced Free Space Optics (FSO): A Systems Approach*, Springer Science+Business Media, New York 2015.
3. Larry C. Andrews and Ronald L. Phillips, *Laser Beam Propagation through Random Media*, Second Edition, SPIE, Bellingham, Washington (2005).

4. V.I.Tatarskii, *The Effects of the Turbulent Atmosphere on Wave Propagation*, Available from U.S. Department of Commerce, Springfield, VA, 22151, 1971. Translated by IPST Staff.
5. D. L. Fried, "Scintillation of a ground-to-space laser illuminator," *J. Opt. Soc. Am.* 57, 980-983 (1967).
6. P.O. Minott, "Scintillation in an earth-to-space propagation path," *J. Opt. Soc. Am.* 62, 685-888 (1972).
7. L.C. Andrews, R. L. Phillips, and P. T. Yu, "Optical scintillation and fade statistics for a satellite-communication system," *Appl. Opt.*. 34,7742-7751 (1995); "Optical scintillations and fade statistics for a satellite-communication system: Errata," *Appl. Opt.* 36, 6068 (1997).
8. R. K. Tyson, "Adaptive optics and ground-to-space laser communications," *Appl. Opt.* 35, 3640-3646 (1996).
9. I. Toselli, "Introducing the concept of anisotropy at different scales for modeling optical turbulence," J.Opt.Soc.Amer. A31(8) (2014) 1868–1875.
10. A. Zilberman, E. Golbraikh, N.S. Kopeika, Propagation of electromagnetic waves in Kolmogorov and non-Kolmogorov atmospheric turbulence: three-layer altitude model, Appl. Opt.47(34)(2008)6385–6391.
11. Xin Shana, Curtis Menyuk, Jing Chen,Yong Ai, "Scintillation index analysis of an optical wave propagating through the moderate-to-strong turbulence in satellite communication links," Optics Communications 445 (2019) 255-261.
12. Hong Guo, Bin Luo, Yongxiong Ren, Sinan Zhao, and Anhong Dang, "Influence of beam wander on uplink ground-to-satellite laser communication and optimization for transmitter beam radius," OPTICS LETTERS, Vol. 35, No. 12, June 15, 2010.
13. Allen Panahi and Alex A. Kazemi, "Optical Laser Cross-Link in Space-Based Systems used for Satellite Communications," Proc. SPIE, Vol. 7675, 76750N-1, 2010.
14. Walter R. LEEB, "Space Laser Communications: Systems, Technologies, and Applications," (Find the Citation..?) The Review of Laser Engineering, 28 (12): 804-808, January 2000. DOI: https://doi.org/10.2184/lsj.28.804
15. Mikhail A. Vorontsov, Svetlana L. Lachinova and Arun K. Majumdar, "Target-in-the-loop remote sensing of laser beam and atmospheric turbulence characteristics," Applied Optica, Vol 55, No. 23, June 2016.
16. J. H. Shapiro, Reciprocity of the turbulent atmosphere, J. Opt. Soc. Am. **61**, 492–495 (1971).
17. A. K. Majumdar and J. C. Ricklin, "Effects of the atmospheric channel on free-space laser communications," in: D. G. Voelz, J. C. Ricklins (Eds.), Free-Space Laser Communications V., Proc. SPIE., Vol.5892, SPIE, Bellingham, WA, 2005, 58920K-1-58920K-16.
18. Arun K. Majumdar and Jennifer C. Ricklin, *Free-Space Laser Communications: Principles and Advances*, Springer, New York 2008.
19. J. Císařaa, O. Wilferta, F. Fanjul-Vélezb, N. Ortega-Quijanob, J. L. Arce-Diegob, "New trends in laser satellite communications: design and limitations," Proceedings of SPIE - The International Society for Optical Engineering. November 2008. DOI: https://doi.org/10.1117/12.818089.

Chapter 4
Optical Laser Links in Space-Based Systems for Global Communications Network Architecture: Space/Aerial, Terrestrial, and Underwater Platforms

4.1 Introduction

The goal of a free-space optical (FSO) communication system is to transmit and receive information efficiently reliably keeping complexity and cost in mind to accomplish the objective. The demand for tremendous high-data-rate and high-bandwidth wireless communication is driven by the enormous increased number of wireless broadband Internet, mobile smartphones, and interactive devices using social networks like YouTube and Facebook for web search, video streaming, and online gaming. FSO wireless communication technology is a potential solution and has the capability to meet the user's ever-increasing demand for bandwidth. FSO wireless communication systems can provide data rates in multi-Gbit/s or higher over typical spans of a few meters to hundreds and thousands of kilometers for a number of applications including:

- Last-mile access network to solve the bandwidth bottlenecks within access network
- LAN-to-LAN interconnectivity to provide high-speed, flexible, and secure connectivity for campus, hospital, and other metropolitan applications
- Optical satellite-to-ground stations as well as constellation satellites to cover remote locations and establish global Internet connectivity
- High-altitude platforms (HAPs)/UAVs to ground station communications
- Inter-satellite communications (e.g., GEO-LEO, LEO-GEO, GEO-GEO links), HAP-LEO, HAP-HAP, etc.
- Underwater communication networks including space/airborne to underwater terminals (e.g., satellite-to-submarine communication)

Satellites or spacecrafts acting as communication nodes connecting with laser/optical links are therefore essentially optical satellite/spacecraft network. Recent successful demonstrations of establishing laser/optical communication links between a satellite and a ground station have opened up the potential of future

© Springer Nature Switzerland AG 2022
A. K. Majumdar, *Laser Communication with Constellation Satellites, UAVs, HAPs and Balloons*, https://doi.org/10.1007/978-3-031-03972-0_4

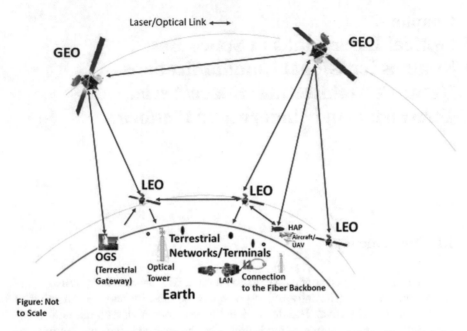

Fig. 4.1 Overview of global optical network integrating satellites, spacecrafts, and terrestrial platforms. (Figure: Not to scale)

high-data-rate communication globally as well as in almost any remote locations. This chapter discusses the development of high-speed (10s of Gbit/s or more) optical networks by integrating space (conventional satellite, small and microsatellites, high-altitude platforms (HAPs), UAVs, balloons, airborne), terrestrial, and underwater platforms. The global communication connectivity and Internet access will thus be possible if the airborne satellites and the fixed terrestrial terminals can all be integrated with appropriate optical devices and laser/optical transceivers.

Figure 4.1 shows an overview of global optical network integrating satellites, spacecrafts, and terrestrial platforms. Figure 4.2a illustrates an overview of global optical network integrating satellites, spacecrafts, and terrestrial platforms, Fig. 4.2b shows the similar optical space network including the *underwater* terminals connecting the space/terrestrial: both above-water and undersea nodes.

a.

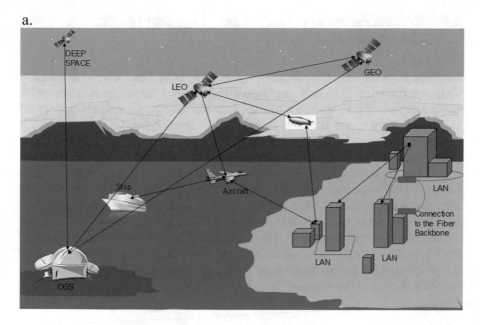

b.

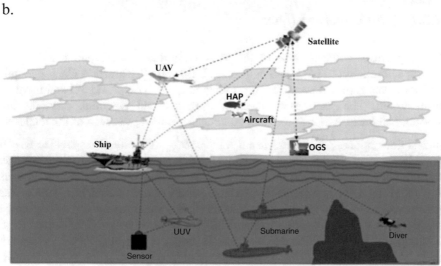

Fig. 4.2 Illustration of the global network integrating space and terrestrial platforms: (**a**) shows an overview of global optical network integrating satellites, spacecrafts, and terrestrial platforms. (*Reprinted with permission from publisher Elsevier, Amsterdam, Netherlands, 2021* [3]); (**b**) shows the similar optical space network including the *underwater* terminals connecting the space/terrestrial: both above-water and undersea nodes. (*Reprinted with permission from Springer Nature and Copyright Clearance Center, 2021* [4])

4.2   Critical Issues and Challenges of Optical Wireless Communication System

Free-space optical wireless communication with large bandwidth can establish high-data-rate space links such as ground-to-satellite/satellite-to-ground, terrestrial, and underwater links (between underwater terminals and satellite-to-underwater/underwater-to-satellite). However, despite great potential, the FSO communication system performance is severely limited mainly by (1) line-of-sight (LOS) requirement (due to a very narrow beam divergence, LOS is a critical requirement for a successful communication), (2) atmospheric turbulence and scattering effects because of the interactions of optical wave with the atmosphere at this short optical wavelength regime, and (3) pointing and tracking requirement in case of ground-to-satellite, inter-satellite, and satellite-to-ground FSO communication links and even for satellite-to-submarine link (which can include both space and random air-water interface). All of these critical issues may result in failure or severe degradation of the performance of the communication system and hence need to be addressed. This section will discuss these issues.

4.2.1   Atmospheric Effects

In the previous chapter, some of the parameters due to atmospheric turbulence effects and relevant to FSO communication links were discussed. A Gaussian optical beam was assumed because it represents most of the typical FSO communication system. The parameters relevant to uplink and downlink channels are scintillation index $\sigma_I^2(r, L)$, spatial coherence radius r_0, variance of the angle-of-arrival fluctuations $\langle \beta_a^2 \rangle$, RMS beam wander σ_{BW}^2, and effective spot size W_{LT}. It was also pointed out that the uplink and downlink parameters are different because of different types of interactions with atmospheric turbulence blobs of various sizes for uplink than for downlink. Furthermore, non-isotropic turbulence, which was also non-Kolmogorov model, was also discussed since reality for a long-range slant-path, optical propagation models that do not necessarily follow conventional Kolmogorov models. Ultimately, an accurate model of optical propagation through atmospheric turbulence is essential for evaluating communication system performance correctly for designing a system properly. For underwater propagation of laser/optical wave, effects of water turbulence on laser propagation are not widely published in the literature. For underwater communication, absorption and scattering losses due to water are primarily important in determining the underwater communication system performance. Absorption and scattering losses due to atmosphere and underwater channels are discussed below.

Absorption and scattering due to particulate matter in the atmosphere may significantly decrease the transmitted optical signal, while turbulence-induced intensity fluctuations can severely degrade the wave-front quality of a signal-carrying laser beam, causing signal fading and random signal losses at the receiver. Optical extinction is caused by absorption and scattering by molecules and aerosol particles predicted by different models. Fog interference at the FSO wavelengths can be significant because of scattering process. Attenuation of an FSO signal is caused by two main factors: (1) absorption and (2) scattering.

4.2.1.1   Atmospheric Attenuation of Laser Power

The transmittance $T(s)$ at frequency v is given by

$$T_v\left(s\right) = \frac{I_v\left(s\right)}{I_{v,0}} = e^{-\beta_v s} \tag{4.1}$$

where the total extinction coefficient β_v comprises the aerosol scattering, aerosol absorption, molecular scattering, and molecular absorption terms as shown below:

$$\beta_v = \beta_{sca,v}^{aer} + \beta_{abs,v}^{aer} + \beta_{sca,v}^{mol} + \beta_{abs,v}^{mol} \tag{4.2}$$

For a slant path, the atmospheric transmittance at a zenith angle, θ, is expressed as β:

$$T_{atm}\left(\lambda\right) = \exp\left[-\sec\theta \int_0^H \beta\left(\lambda, h\right) dh\right] \tag{4.3}$$

where β is the total atmospheric extinction coefficient, H is the vertical height of the atmospheric channel, and λ is the FSO communication wavelength.

The total extinction coefficient is expressed in km^{-1}. The signal loss in the link budget calculation L_a due to atmospheric extinction can be computed by taking the logarithm of transmittance to the base 10 and multiplying by 10 and is expressed in dB:

$$L_a = 10\log_{10}\left[T\left(s\right)\right] \tag{4.4}$$

Some of the typical values of the total extinction coefficient are clear air $\beta_v \approx 0.1$; haze conditions (with visibility of roughly 4 km) $\beta_v \approx 1.0$; and in fog conditions, $\beta_v \geq 10$, and for dense fog with a visibility of 10 m, $\beta_v \approx 391$. The loss for a typical hazy condition with $\beta_v = 1.0$ at a wavelength of about 1.55 µm and for a 2 km path is $L_a = 10\log_{10}(e^{-2}) \approx 8.7$ dB.

Atmospheric conditions responsible for attenuation of FSO communication signal are fog, snow, rain, and turbulence [1]. Table 4.1 shows the all-weather broadband availability under various visibility conditions (m), attenuation in dB/km at 785 nm, and FSO system range (m) with precipitation in mm/h for different weather conditions.

Table 4.1 All-weather broadband availability for different weather conditions [2]

Weather condition	Precipitation	mm/hr	Visibility	dB/km Loss (785 nm)	FSO System Range
Dense fog			0 m		0 m
			50 m	-339.6	110 m
Thick fog			200 m	-84.9	330 m
Moderate fog	Snow		500 m	-34.0	610 m
Light fog	Cloudburst	100	770 m	-20.0	780 m
			1 km	-14.2	880 m
Thin fog	Heavy rain	25	1.9 km	-7.1	1.02 km
			2 km	-6.7	1.03 km
Haze	Medium rain	12.5	2.8 km	-4.6	1.08 km
			4 km	-3.0	1.13 km
Light Haze	Light rain	2.5	5.9 km	-1.8	1.18 km
			10 km	-1.1	1.22 km
Clear	Drizzle	0.25	18.1 km	-0.6	1.25 km
			20 km	-0.53	1.26 km
Very Clear			23 km	-0.46	1.26 km
			50 km	-0.21	1.28 km

In free-space communication, the so-called last-mile problem (or last-mile bottleneck) relates to connecting the high bandwidth from the fiber-optic backbone to all of the business with high-bandwidth networks. Less than 5% of most of the buildings in the USA have a direct connection to the very-high-speed (~2.5–19 Gbit/s) fiber-optic backbone, yet more than 75% of the businesses are within 1 mile of the fiber backbone. An architecture showing "*small cell*" concept to develop basic technology block to achieve *all-weather 99.999%* ("*five nines*") *availability* is reported [3, Ch. 3] based on a link budget analysis and visibility-limiting weather conditions (even in the worst case of dense fog!) indicating to meet carrier-class availability (99.999%: "five nines"). For FSO communication signal at optical wavelength, fog is the worst effect due to scattering effects which can be even as high as 350 dB/km. Rain and snow have less adverse effects for optical links and can be in the range of 45 and 150 dB/km in heavy rain or snow [2]. The researchers showed [2] that a range of 140 m is still possible even in the presence of worst case of 350 dB/km using a transmitter power of 30 mW and a detector of sensitivity 25 nW.

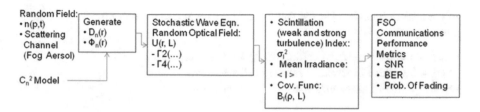

Fig. 4.3 A simplified flowchart showing how the FSO communications performance metrics are derived

4.2.2   Free-Space Optical (FSO) Communication System Performance: Terrestrial and Satellite/Airborne Platforms

In order to design a free-space optical communication system properly, it is essential to understand the propagation effects which ultimately limit the communication system performance such as achievable data rates and communication capacity. A number of recent applications for providing high-bandwidth wireless communication links include satellite-to-satellite cross-links and uplink and downlink between satellite, aircraft, HAP, UAV, and optical ground platforms, and among mobile or stationary terminals to address the current 5G/6G and IoT issues. However, there are a number of deleterious features of the atmospheric channel effects that may lead to serious signal fading, and even a complete loss of signal altogether. Intensity fading and random signal losses at the receiver may be due to absorption/scattering effects as well as random turbulence effects that severely degrade the wave-front quality of a signal-carrying laser beam. From communication system performance point of view, it is therefore necessary to analyze some of the parameters such as signal-to-noise ratio (SNR) at the receiver, achievable data rate, bit error rate (BER) of the digital information, and FSO link reliability and probability of fade. Most of these are already discussed [3, 4], and interested readers may look at the appropriate chapters.

One of the essential parameters in determining the FSO communication system performance is scintillation index:

$$\sigma_I^2 (r, L) = \frac{\langle I^2 (r, L) \rangle}{\langle I (r, L) \rangle^2} - 1 \tag{4.5}$$

Scintillation index is the normalized variance of optical intensity fluctuations with respect to the mean irradiance. For a transmitting Gaussian beam, the scintillation index expressions for weak and strong turbulence are discussed [5]. The step-by-step processes to derive the necessary statistical quantities from random refractive index parameters to finally arrive at the necessary system parameters for designing the FSO communication systems are shown in Fig. 4.3. The symbols used in the system blocks [3] are as follows and are related to the solutions of the optical wave equation [4]:

$n(r, t)$ = Random index of refraction (spatial and temporal)
$D_n(r)$ = Structure function of random refractive index
$\Phi_n(r)$ = Three-dimensional spatial power spectrum

4.2.2.1   Probability Density Function (PDF) Models for Irradiance Fluctuations Relevant to FSO Communication System Performance

The performance of a free-space lasercom system is reduced by irradiance scintillations due to amplitude and phase fluctuations of the optical propagation through atmospheric turbulence. All these effects finally degrade FSO system performance for both terrestrial and slant paths (uplink and downlink) even in clear air condition. An accurate knowledge of the PDF for the received signal under all intensity fluctuation regimes is difficult. An acceptable PDF model is essential to determine the reliability of FSO communication system, which depends on the probability of detection, miss and false alarm rate, and probability of fade. Various statistical models for irradiance PDF have been proposed, which are valid for both weak and strong turbulence regime. The two mostly commonly used PDFs are log-normal and gamma-gamma, where log-normal distribution is applicable to weak turbulence case and gamma-gamma is generally valid for both weak and strong turbulence regimes [3–5].

4.2.2.2   Received Signal-to-Noise Ratio (SNR) and Bit Error Rate (BER)

The ultimate limit to the detectability of a weak signal is sent by *noise*, which is an unwanted signal that obscures the desired signal. For FSO communication system, the shot noise, background noise, and thermal noise contribute to the total noise at the receiver. For a digital communication system, the goal is to transmit the maximum number of bits per second over the maximum possible range with the fewest errors at the receiver. Communication capacity for a digital communication system to deliver high data rate depends on this concept. At the transmitter end, electrical data signals are converted to optical signals via a modulator where a "1" is transmitted as a pulse light while "0" has no light output. The bit rate is determined by the number of "1s" and "0s" transmitted per second. The optical signal is detected at the receiving end by a photodetector, where a signal-processing unit with a decision circuit identifies the "1s" and "0s" in the signal to receive the message. Information for a digital communication system is sent over an optical link as digital symbols. The source information is encoded into binary symbols (bits), and the bits are transmitted as some type of coded optical field; for example, it can be encoded on a bit-by-bit basis (binary encoding). Each bit is then sent individually by transmitting one of the two optical fields to represent each bit. At the receiver, decoding is based on a decision to determine the best performance in decoding the correct signal with the lowest probability of making a bit decision error, and thus the bit error rate (BER) can be obtained.

4.2.2.3   Relationship Between SNR and BER

FSO communication system performance can be evaluated by computing the bit error rate (BER) of the system, which depends on the particular modulation format, and the received signal-to-noise ratio (SNR). The effective noise comes from the following sources: signal shot noise, dark current noise, thermal/Johnson noise in the electronics following the photodetector, and background noise. Assuming a Gaussian distribution of noise, the SNR at the output of the photodetector is given by [3]

$$SNR_0 = \frac{P_S}{\sqrt{\left(\frac{2hvB}{\eta}\right)(P_S + P_B) + \left(\frac{hv}{\eta e}\right)^2 \left(\frac{4kT_N B}{R}\right)}} \tag{4.6}$$

where P_S is the signal power (Watts) of the optical transmitter, P_B is the background noise (Watts), η is the detector quantum efficiency, e is the electronic charge in coulomb, h is the Planck's constant, v is the optical frequency (Hz), k is the Boltzmann constant, B is the detector filter bandwidth, T_N is the effective noise temperature, and R is the effective input resistance to the amplifier of the detector. For shot noise-limited system, the background noise and thermal noise can be neglected. Note that the SNR is an instantaneous value of the signal-to-noise ratio. In the presence of atmospheric turbulence, the SNR is a fluctuating term, and therefore the average (mean) value has to be taken. The mean SNR can be written as

$$\langle SNR \rangle = \frac{SNR_0}{\sqrt{\frac{P_{S0}}{\langle P_S \rangle} + \sigma_I^2(D)SNR_0^2}} \tag{4.7}$$

where SNR_0 is the signal-to-noise ratio (SNR) in the absence of turbulence; P_{S0} is the signal power in the absence of atmospheric effects; $\langle P_S \rangle$ is the mean input signal power, i.e., the mean of the instantaneous input is signal power, P_S; and $\sigma_I^2(D)$ is the aperture-averaged scintillation index which depends on the receiver aperture diameter, D.

The communication system performance parameter bit error rate (BER) is the probability of incorrect bit identification by the decision circuit and is an important performance figure of merit for communication system design. The FSO communication system receiver sensitivity is defined as the minimum averaged received optical power required to achieve a given BER, which is one of the design requirements. The probability of error in the absence of turbulence is given by (assuming a Gaussian distribution for both the noise and signal plus noise)

$$\mathrm{BER}_0 = \frac{1}{2}\left(\frac{\mathrm{SNR}_0}{2\sqrt{2}}\right) \tag{4.8}$$

In the presence of turbulence, the probability of error can be written as [3]

$$\mathrm{BER} = \frac{1}{2}\int_0^\infty p_\mathrm{I}(s)\,\mathrm{erfc}\left(\frac{\langle \mathrm{SNR}\rangle s}{2\sqrt{2}\langle i_\mathrm{S}\rangle}\right)ds \tag{4.9}$$

where $p_\mathrm{I}(s)$ is the probability distribution of irradiance, i_S is the instantaneous signal current with a mean value $\langle i_S\rangle = \eta e\langle P_S\rangle/h\nu$, $\langle P_S\rangle$ is the mean signal power, and $\langle \mathrm{SNR}\rangle$ is the mean SNR in the presence of turbulence. The PDF of intensity fluctuations $p_\mathrm{I}(s)$ can be taken from the accepted model, and a gamma-gamma probability density function is generally accepted for both weak and strong turbulence cases. Some typical values of BER are in the ranges of 10^{-6} to 10^{-9} for design purposes.

4.2.2.4   FSO Communication Signal Temporal Frequency Spectrum

Temporal power spectrum of irradiance fluctuations is important since a quantitative analysis of the temporal behavior of the irradiance provides significant merit of figures in the optical link of an FSO communication system. Instantaneous values of the signal-to-noise ratio (SNR) are also important in order to evaluate the burst error rate and the overall availability of the FSO communication system [6]. The temporal characteristics of an FSO system are essential in developing and designing optimal schemes for detection and coding. Also see the book by H.R. Anderson [26]. The temporal spectrum of irradiance fluctuations, or the power spectral density (PSD), $S(\omega)$, by the Fourier transform of the temporal covariance function is given by

$$S(\omega) = 2\int_{-\infty}^{\infty} B_\mathrm{I}(\tau, L)\mathrm{e}^{-j\omega\tau}d\tau = 4\int_{-\infty}^{\infty} B_\mathrm{I}(\tau, L)\cos(\omega\tau)d\tau \tag{4.10}$$

where $B_\mathrm{I}(\tau, L)$ is the temporal covariance function. Recently, the temporal coherence time is computed from the inverse of frequency spread for a specific wind velocity of 10 km/h and for two strengths of turbulence, $C_n^2 = 1.8\times10^{-13}$ m$^{-2/3}$ and $C_n^2 = 1.8\times10^{-15}$ m$^{-2/3}$. The result [7] is shown in Fig. 4.4.

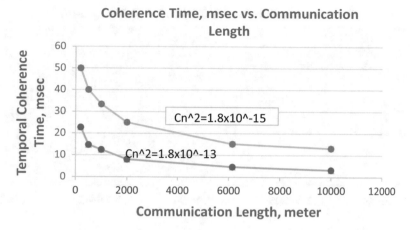

Fig. 4.4 Temporal coherence time versus communication range for two strengths of turbulence and for a wind velocity of 10 km/h

Atmospheric turbulence can degrade the performance significantly, and the temporal diversity can be exploited to improve the performance of the transmission link. Depending on the acceptable delay latency and some degree of time diversity, one can employ various channel coding so that the system performance can be improved. Figure 4.5 depicts the results of both weak and strong turbulence and shows the performance improvements at a typical value of BER = 10^{-5} with respect to the no-coding case (i.e., improved coding gain, in dB) for two types of coding: RSC4 and TC with "potential time diversity order (PTDO)" of 2 [7]. Note that the coding gains depend on both acceptable BER and strength of turbulence. A gamma-gamma probability density function (PDF) of intensity fluctuations is assumed in this analysis.

4.2.2.5   The Outage Probability Due to Fading Relevant to FSO Communication Performance

In designing an FSO communication system, it is absolutely important to obtain uninterrupted communication between the transmitter and the receiver during the exchange of information sending and receiving data under all atmospheric conditions. But because of the presence of randomly varying atmospheric channel, the received signal fluctuates and at times can even fall below an acceptable detection level and error rate. The fading probability of an optical communication signal can be computed from a knowledge of statistics of intensity fluctuations (PDF) and scintillation index. The reliability of an FSO communication system thus depends on the fading probability. The probability of fade can then be computed from the fraction of times the instantaneous value of irradiance $I(t)$, normalized to its mean, exceeds the threshold level of intensity I_T specified by the communication system

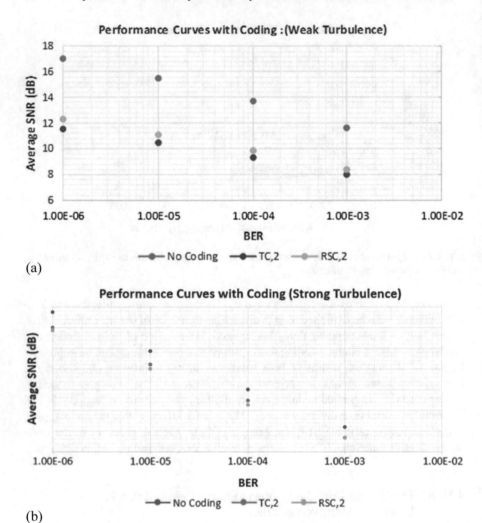

Fig. 4.5 Communication performance curves with coding for (**a**) weak turbulence and (**b**) strong turbulence

requirement, i.e., $I(t) \geq I_T$. Since the statistics of optical propagation can be assumed as an ergodic process in which case the time averages are equal to ensemble averages, we can write the fraction of time of a fade where $I \leq I_T$ in the following way:

$$\text{Fraction } (I \leq I_T) = \text{CP}_I(I \leq I_T)$$

$$= \int_0^{I_T} p_I(I) \, dI \qquad (4.11)$$

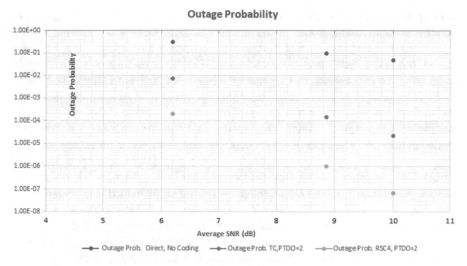

Fig. 4.6 Outage probability for direct link and link with coding

where CP_I is the cumulative probability for irradiance (e.g., log-normal or gamma-gamma distribution). The fade threshold parameter F_T of a signal in dB can be defined by

$$F_T = 10\log_{10}\left(\langle I(0,L)\rangle / I_T\right)$$

where $\langle I(0,L)\rangle$ is the on-axis mean intensity and L is the communication range. The probability of fade as a function of the threshold parameter F_T for various communication scenarios, such as downlink, uplink, and horizontal links for two PDF models of log-normal and gamma-gamma distributions, is reported [3, 4]. Figure 4.6 shows the examples of the outage probability as a function of available SNR for direct link and links with coding [7]. The analysis of outage probability is shown as a function of average SNR (dB) for (1) no coding; (2) TC, PTDO = 2; and (3) RSC4, PTDO = 2.

Note that outage probability depends significantly on the coding chosen and depends on the average SNR.

4.2.2.6   Modulated Retroreflector-Based FSO Communication Concept

Another communication system concept of retroreflective communication system will also be addressed in the space/underwater laser communication system. This is based on data links using modulated retroreflectors or MRR. This communication technology of data transfer can support "shutters" that can achieve usable communication data rates in Gbit/s ranges. MRR require very little power draw and offer extremely small form factors and mass, which will be very practical to install on underwater vehicles and almost any space-based platform. The MRR communication system can be designed to minimize the challenge of beam alignment and deployment. Some of the features of retro-modulators include remote optical communications at a high data rate between a base station (can be space based and underwater) and a remote station (can also be underwater and space based), which can consist of two sets of lenslets coupled with single-mode fiber array (fiber retro) and can operate under atmospheric and underwater channels. The technology can thus provide a means of extremely fast remote data transfer with additional advantage of maintaining secure data transfer (recovery) achieving a wide field of view. The technology concept has been described in recent reports [4, 8] and can be easily applied to space/underwater duplex laser communications. The effects of atmospheric turbulence on the amplified fiber-retro-modulator are discussed [4], and BER of the communication system performance is analyzed in the presence of atmospheric turbulence and scattering. An MRR-equipped FSO communication system operating as duplex performing two-way optical communications is shown in Fig. 4.7. The system concept can be used for both space/air and between underwater platforms for collecting and recovering (remotely probing) information for both identifying and communication purposes. For example, one of the potential scenarios may be that an airborne, UAV, or even small/nanosatellite (equipped with an MRR) can collect the intelligence information of underwater target with a camera or an appropriate sensor system placed on the platform.

A LEO satellite can receive (recover) the information by simply sending a continuous-wave (CW) signal to the MRR. It is interesting to note that MRR can also complement LADAR system placed on a space/underwater platform for location and identification of underwater targets of interest.

4.2.3   Underwater Free-Space Optical Communications (uFSO)

Underwater free-space optical (uFSO) communications have recently gained a considerable interest for developing communication systems between underwater platforms and underwater/above-water terminals. Scattering and variable optical qualities of ocean water need to be considered since these varying properties change with time and location, all of which affect the amount of light lost. There are numerous similarities between uFSO and free-space optical (FSO) communications of laser satellite links due to the fact that both use optical wavelengths to transfer high-data-rate secure information between line-of-sight (LOS) links with small size, weight, and power (SWaP) components.

4.2.3.1   Designing an Underwater FSO Communication (uFSO) System

There is an urgent need to develop underwater high-speed communication because of a number of applications such as monitoring of deep-sea oil by autonomous underwater vehicle (AUV), which could be stationed at the cell and communicate

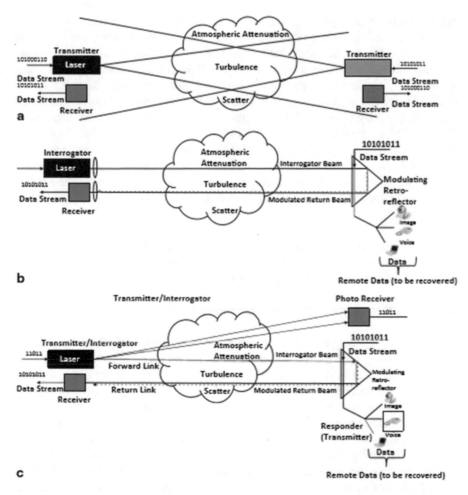

Fig. 4.7 An MRR-equipped FSO duplex communication system concept for applications both in space/air and between underwater platforms. (**a**) Conventional FSO communication system. (**b**) MRR-based FSO communication system (half-duplex, HDX mode). (**c**) MRR-based FSO communication system concept (full duplex, FDX mode). *FSO* free-space optical, *MRR* modulated retroreflector. (*Reprinted with permission from Springer Nature and Copyright Clearance Center, 2021* [4])

surveillance data wirelessly by AUVs to communicate with a base station. Another important application is to establish a broadband global communication and Internet connectivity using satellite, terrestrial, and underwater platforms with space, aerial, and underwater communication links. Blue-green laser can be used for maximum penetration in seawater, and blue-green optical communication system helps to establish a communication between a space or aircraft and submerged objects. This is possible because blue-green waves exhibit much lower attenuation (scattering and absorption) when propagated through water and therefore are most suitable for underwater communication. Most of the laser cannot penetrate through sea because the beam is absorbed by the sea and the blue-green laser in the region of 470–570 nm has the minimum energy fading in the sea, with a fading rate of about 0.155–0.5 dB/m. The blue-green laser can therefore propagate from several hundreds of meters to kilometers in the sea, and some submarine laser communication systems are being developed in this wavelength range. However, blue-green optical communications are generally applicable to shorter distances. Various techniques are being developed to improve the FSO communication between a submerged object and an air platform. Recently, researchers have reported [25] a low-power 520 nm green laser diode (LD)-based underwater communication system and experimentally demonstrated to implement maximal communication capacity of up to 2.70 Gbit/s data rate over a 34.5 m underwater transmission distance. The authors used nonreturn-to-zero on-off keying (NRZ-OOK) modulation scheme and even achieved a maximum data rate of 4.60 Gbit/s for a shorter underwater distance of 2.3 m.

4.2.3.2   Fundamental Physics of All-Underwater and Underwater/ Above-Water Propagation Channel: Impact on Communications Performance

Underwater Optical Wireless Communications (uOWC)

The water as the propagation medium presents absorption in uOWC, which depends on the inherent properties of the seawater and the relative concentrations of algal and non-algal matter. Compared to FSO communication where aerosols, gases, and fogs produce scattering in uOWC, this spatial dispersion arises mainly from phytoplankton. Turbulence due to random variations of atmospheric refractive index is one of the main atmospheric parameters limiting the FSO communication system performance, whereas for uOWC, refractive index of seawater depends on temperature, salinity, and pressure. For underwater optical communications, the random changes in seawater refractive index are not enough to model oceanic turbulence accurately. However, the phenomenon of scintillations caused by turbulence in water has some effects in the uOWC link parameters and will be discussed below. One other factor which also determines the uOWC link parameters is random air-water propagation of optical waves, which is important in underwater-to-air laser link that will also be discussed below.

When optical pulses propagate through an aquatic medium, they suffer from attenuation and broadening in the spatial, angular, temporal, and polarization domains, which ultimately limit the underwater free-space optical (uFSO) communications performance. Attenuation and broadening are wavelength dependent and are caused by multi-scattering of light by water molecules and marine hydrosols.

In order to determine the feasibility and reliability of underwater optical links, variations in the optical properties of ocean water need to be considered. High-speed optical links with high-brightness blue-green LED sources and laser diodes can be used for short-range applications because underwater systems have severe power and size constraints compared to terrestrial or air-based systems. Furthermore, in different optical environments such as shallow water, the effects of absorption by organic matter and scattering by inorganic particulates can be severe compared to deep ocean water. These effects are also different near the seafloor than on the middle of the more homogeneous-type water column. Fluctuating wave surfaces and the background sunlight can also contribute to the received or transmitted optical signals to or from the underwater platforms.

The extinction coefficient $c(\lambda)$ can be expressed as a sum of the absorption coefficient $\alpha(\lambda)$ and scattering coefficient $\beta(\lambda)$ for different types of water including pure seawater, clean ocean water, coastal ocean, and turbid harbor water at the FSO communication wavelength of interest (typically ~450–550 nm range approximately). This can be written as

$$c(\lambda) = \alpha(\lambda) + \beta(\lambda) \tag{4.12}$$

For a propagation distance of z, the propagation loss factor as a function of wavelength is given by the following:

Underwater Attenuation of Laser Power

The underwater transmittance at the optical wavelength λ and range z can then be written as

$$T_{UW}(\lambda, z) = \exp\left[-c(\lambda)z\right] \tag{4.13}$$

Some of the detailed modeling of the underwater absorption, scattering, and attenuation has been recently reported [9]:

Modeling Absorption

This can be modeled as a linear combination of the absorption properties of pure seawater, chlorophyll absorption as a function of wavelength and concentration, and two-component color-dissolved organic material (CDOM):

$$\alpha(\lambda) = \alpha_w(\lambda) + \alpha_{cl}(\lambda) + \alpha_f(\lambda) + \alpha_h(\lambda) \tag{4.14}$$

where $\alpha_w(\lambda)$ is the absorption coefficient of pure water, $\alpha_{cl}(\lambda)$ is the absorption coefficient of chlorophyll, $\alpha_f(\lambda)$ is the fulvic acid absorption coefficient, and $\alpha_h(\lambda)$ is the humic acid absorption coefficient. When all the various components of humic and fulvic acid are added together, the total absorption coefficient can be obtained as follows [9]:

$$\alpha(\lambda) = \alpha_w(\lambda) + \alpha_c^0(\lambda)\left[C_c / C_c^0\right]^{0.0602}$$
$$+ \alpha_h^0 * C_h \exp(-k_h\lambda) + \alpha_f^0 * C_f \exp(-k_f\lambda) \tag{4.15}$$

where the suffixes c, h, and f denote the components of chlorophyll, humic, and fulvic absorption coefficients, respectively. The following values are taken from [9]

$$C_c^0 = 1 \text{ mg} / \text{m}^3, \quad \alpha_h^0 = 18.828 \text{ m}^2 / \text{mg}, \quad k_h = 0.01105 / \text{nm},$$
$$k_f = 0.0189 / \text{nm}, \quad \alpha_f^0 = 35.959 \text{ m}^2 / \text{mg}$$

The parameters C_c, C_h, and C_f are the total concentration of chlorophyll, humic acids, and fulvic acids in mg/m³, respectively.

Modeling Scattering

The total scattering coefficient $\beta(\lambda)$ is a linear combination of the scattering coefficient of pure water, $\beta_w(\lambda)$; the scattering from small particles, $\beta_s^0(\lambda)$; and the scattering from large particles. The total scattering can be written as [6]

$$\beta(\lambda) = \beta_w(\lambda) + \beta_s^0(\lambda) * C_s + \beta_l^0(\lambda) * C_l \tag{4.16}$$

From the values reported in [9], the total scattering is

$$\beta(\lambda) = \frac{0.005826}{m} * \left(\frac{400}{\lambda}\right)^{4.322} + 1.151302 \left(\frac{m^2}{g}\right) * \left(\frac{400}{\lambda}\right)^{1.7}$$
$$* C_s + 0.3411 \left(\frac{m^2}{g}\right) \left(\frac{400}{\lambda}\right)^{0.3} * C_l \tag{4.17}$$

Total Attenuation Modeling

The total attenuation is the sum of both absorption coefficient and scattering coefficient and therefore can be written as

$$c(\lambda) = \alpha(\lambda) + \beta(\lambda)$$

where the detailed values of the absorption and scattering coefficient are expressed above.

Turbulence Effect in Oceanic Water

Optical wireless communications are affected by the optical turbulence generated by random spatial and temporal microscale fluctuations of optical refractive index. The major effect of optical turbulence on free-space optical communications is the intensity fluctuations at the received signal and is generally termed as scintillation. Scintillation effects from turbulent water arise because of the variations of the refractive index due to variations from flow, salinity, and temperature as well as due to stratified layers of water. Compared to the effects due to absorption and scattering, turbulence can be considered to be a secondary effect and is therefore less important. However, it is still important to evaluate the scintillations of underwater optical signals because in addition to absorption and scattering effects, discussed earlier in this section, it can also limit the available underwater data rate using optical links.

Optical turbulence is due to rapid change of the seawater refractive index occurring at any depth. The power spectrum of the refractive index in seawater turbulence depends on both temperature and salinity and is therefore different from the so-called Kolmogorov power spectra widely applicable to FSO communication links.

For a Gaussian beam propagating through weak turbulence and without taking into account of scattering effects, scintillation index $\sigma_I^2(\bar{r},L,\lambda)$ can be expressed [10] as a sum of both longitudinal and radial components:

$$\sigma_I^2(\bar{r},L,\lambda) = \sigma_{I,l}^2(0,L,\lambda) + \sigma_{I,r}^2(\bar{r},L,\lambda) \qquad (4.18)$$

where λ is the optical wavelength, $k = (2\pi/\lambda)$ is the wave number, L is the propagation distance, and κ is the spatial frequency (inverse of wavelength),

The longitudinal component is

$$\sigma_{I,l}^2(0, L, \lambda) = 8\pi^2 k^2 L \int_0^1 \int_0^\infty \kappa \, \Phi_n(\kappa) \exp\left(-\frac{\Lambda L \kappa^2 \xi^2}{k}\right)$$
$$\times \left[1 - \cos\left(\frac{L\kappa^2}{k} \xi\left(1 - (1-\Theta)\xi\right)\right)\right] d\kappa d\xi$$

and the radial component is

$$\sigma_{I,r}^2(\bar{r}, L, \lambda) = 8\pi^2 k^2 L \int_0^1 \int_0^\infty \kappa \, \Phi_n(\kappa) \exp\left(-\frac{\Lambda L \kappa^2 \xi^2}{k}\right) \left[I_0\left(2\Lambda r\xi\kappa\right) - 1\right] d\kappa d\xi \quad (4.19)$$

$I_0(x)$ is the zeroth-order Bessel function, Λ and Θ are the Gaussian beam parameters (see also [5]), W_0 is the radius of the Gaussian beam at the $1/e$ point, and F_0 is the radius of curvature of the beam front:

$$\Lambda = \frac{\dfrac{2L}{kW_0^2}}{\left(\dfrac{2L}{kW_0^2}\right)^2 + \left(1 - \dfrac{L}{F_0}\right)^2}$$

$$\Theta = \frac{1 - \dfrac{L}{F_0}}{\left(\dfrac{2L}{kW_0^2}\right)^2 + \left(1 - \dfrac{L}{F_0}\right)^2}$$

$\Phi_n(\kappa)$ is the power spectrum of turbulence. For homogeneous and isotropic oceanic waters, the power spectrum is given by [10–12]

$$\Phi_n(k) = 0.38 \cdot 10^{-8} \varepsilon^{-\frac{1}{3}} \kappa^{-\frac{11}{3}} \left[1 + 2.35(\kappa\eta)^{\frac{2}{3}} \right] \frac{\chi_T}{w^2} \left(w^2 e^{-A_Y\delta} + e^{-A_S\delta} - 2we^{-A_{TS}\delta} \right) \quad (4.20)$$

In the above equation, $\eta = 10^{-3}$ m is the Kolmogorov microscale, ε is the rate of dissipation of turbulent kinetic energy per unit mass of fluid and ranges from 10^{-8} to 10^{-2} m²/s³, and χ_T is the rate of dissipation of mean square temperature, ranging from 10^{-10} to 10^{-4} K²/s:

$$\delta = 8.284(\kappa\eta)^{4/3} + 12.978(\kappa\eta)^2 \text{ and }$$
$$A_T = 1.863 \times 10^{-2}, \ A_S = 1.9 \times 10^{-4} \text{ and } A_{TS} = 9.41 \times 10^{-3}$$

w is a unitless variable specifying the relative strength of the fluctuations caused by temperature or salinity in the seawater. Strong oceanic turbulence (i.e., when Rytov variance $\sigma_R^2 \gg 1$) can occur at a shorter distance than 100 m. This can therefore limit the available data rate for underwater optical communication performance. For weak turbulence regime of underwater scintillation, a log-normal probability density function of intensity fluctuations can be assumed. For strong turbulence regime which can occur for a longer distance underwater optical path link, the combination of exponential and log-normal distribution, a two-lobe distribution is proposed by Jamali et al. [13] for received underwater signal in the presence of air bubbles. The PDF of the two-lobe distribution is given by [13]

$$f_{\tilde{h}}(\tilde{h}) = \frac{k}{\gamma} \exp\left(-\frac{\tilde{h}}{\gamma} \right) + \frac{(1-k)\exp\left(-\frac{\left(\ln \tilde{h}\mu \right)^2}{2\sigma^2} \right)}{\tilde{h}\sqrt{2\pi\sigma^2}} \quad (4.21)$$

where the parameter k determines the balance between the two distributions, γ is the mean of exponential distribution, and μ and σ^2 are the mean and variance of the log-normal distribution, respectively.

Random Air-Water Interface Relevant to uOWC

This subject was discussed in an earlier chapter. The essential mathematical formulations of propagation through the random wavy air-water interface from the earlier chapter are summarized below.

For water surface *reflection* case, the probability density function (PDF) of the angular fluctuations of an emergent angle is given by

$$PDF_D\left(\theta_D\right) = \frac{1}{2}PDF_S\left(\frac{\theta_D}{2}\right) \tag{4.22}$$

where θ_D = angle by which this reflection deviates from its position in the case of a calm, flat surface, and $PDF_S(\theta_S)$ is the PDF for the wave slope θ_S of the time-varying surface at the moment of contact.

For the transmission through random air-water interface, using the simplest estimate for the beam wander PDF, the PDF is given by

$$PDF_D\left(\theta_D\right) \cong 4PDF_S\left(-4\theta_D\right) \tag{4.23}$$

where $\theta_D \cong -\frac{1}{4}\theta_S + O(2)$.

Using these above results, the beam wander effects applicable to both reflection and transmission cases and the uFSO communication parameters such as BER as a function of SNR for both reflected and transmitted are shown in the report [14].

Impulse Response for uOWC

Communication channel response is very critical to predict the channel gain and bandwidth (also maximum achievable data rate). Predicting the model depends on the physical properties of the medium and the typical geometrical parameters of a particular underwater link scenario. Channel response modeling for uOWC shallow-water links is the most difficult because it is the most variable due to high presence of chlorophyll, particles, fauna, random nature of the wind-driven seawater, and effect of sunlight. All of these factors limit the available received optical signal, which determines the communication system performance. The impulse response effect is more stable for a deeper link. An estimation of the impulse of response for uOWC links is reported [15]. The reported research [15] investigates the pulse propagation and solves the frequency-domain vector radiative transfer equation for a total calculated bandwidth of 6.4 GHz, which is sufficient for the 1 Gbit/s uOWC study of accomplishing data rate. His theoretical model results show the impulse responses for a 30 m (optical depth = 9.1641) and 50 m (optical depth = 15.2735) underwater link. The results show that the peak of the co-polarized impulse responses at −40 and −66 dB are for 30 and 50 m underwater links, respectively, with very *sharp* rise from 0.13 to 0.133 µs for 30 m and *sharp* rise from 0.22 to 0.222 µs for 50 m link. The *slower* impulse response fall occurs from 0.133 to 0.18 µs for 30 m case and from 0.222 to 0.27 µs for 50 m. For cross-polarized component, the impulse response is reduced approximately by less than 10 dB, which shows the effect of multiple scattering on the increase of the cross-polarized component. The application of polarization to uOWC can thus be evaluated.

Categories of uOWC Links

Considering transmitter and receiver nodes for uOWC network, the following types of optical wireless communication links are useful for various underwater scenarios and requirements: (1) point-to-point line-of-sight (LOS) configuration, (2) retroreflector-based LOS configuration, (3) diffused configuration, and (4) non-line-of-sight (NLOS) configuration. The communication performance decreases drastically for underwater optical wireless communication mainly because of increased water absorption. It is possible to improve the basic water channel absorption effect by using the scattered light, which can be achieved by designing various link configurations. Propagation of light pulses in aquatic medium is affected by attenuation and broadening (stretches) in the spatial, angular, temporal, and polarization domains, which ultimately limit the uOWC performance. The absorption, scattering, and extinction coefficients at 520 nm optical wavelength for four types of water, pure seawater, clean ocean water, coastal ocean water, and turbid water, are discussed [16, 17].

In order to evaluate uOWC system performance, the communication link models for the three link scenarios and the bit error rate (BER) have been reported and are given by [18, 19]. Underwater optical link is established by line-of-sight (LOS) propagation between the transmitter and the receiver located at two underwater terminals.

Received optical signal power:

$$P_{\mathrm{R\_los}} = P_{\mathrm{T}}\eta_{\mathrm{T}}\eta_{\mathrm{R}}L_{\mathrm{pr}}\left(\lambda, \frac{d}{\cos \theta}\right)\frac{A_{\mathrm{Rec}}\cos\theta}{2\pi d^2\left(1-\cos\theta_0\right)} \tag{4.24}$$

where P_{T} is the transmitting optical power with a divergence angle θ_0 (assumed to be very narrow); η_{T} and η_{R} are the optical efficiencies of the transmitter and the receiver, respectively; L_{pr} is the propagation loss factor; d is the distance between the transmitter and the receiver; θ is the angle between perpendicular to the receiver plane and the transmitter-receiver trajectory; and A_{Rec} is the receiver aperture size. In some scenarios, the terminal at the transmitter may communicate with a moving terminal such as a diver equipped with a modulating retroreflector (MRR), which is capable of simultaneously reflecting back the modulating signal whenever the MRR is interrogated by the transmitting signal. This way, a duplex communication can be possible between the transmitter (for example located in a submarine) and a moving target (a diver or another autonomous moving terminal) without the need for accurate pointing, acquisition, and tracking (PAT) system. The underwater communication system parameters and the BER as a function of transmitter-receiver separation for the above different scenarios are discussed [18, 19]. In another communication scenario, there may be an underwater obstruction between the optical transmitter and the receiver LOS propagation path or misalignment of the optical transceiver system so that the only way a transmitting beam can reach the receiver is by reflecting from the ocean-air interface via total internal reflection. From practical point of view, this will be possible if the ocean-air surface is smooth. But as discussed previously, random air-water interface can cause random fluctuations of the received optical beams, which will help to maintain communication link.

4.2.3.3 Underwater/Above-Water Propagation Channel: Impact on Communications Performance

The concept of developing blue-green laser links for establishing satellite-submarine duplex communications has the potential to improve connectivity with underwater and submarine assets. Integrated airborne and spaceborne architectures including downlink, uplink, and duplex communication capabilities will make the concept extremely effective. The integrated optical network should include both space and undersea network.

This section describes system concept, architectures, and communication performance for integrated space/aerial, terrestrial, and underwater FSO communication systems. The ability to achieve communications in integrated FSO communication system is very important in various tactical and strategic communications. One of the objectives to establish this tactical communication requires FSO communication platforms that involve satellite (for example LEO), small/microsatellite, high-altitude platform (HAP), airborne/UAV terminal, optical ground station (OGS), terrestrial gateways, and underwater terminals such as submerged submarine, in depths down to about 100 m. A globally distributed free-space optical (FSO) communication system utilizes multiple geographically separated FSO wireless communication systems to communicate with each other. Distributed optical sensor systems can process data and then transmit it, receive data and then process it, or have some combination of the two. The main challenges of any distributed system are localization, synchronization, and processing capabilities. An FSO link can be utilized for communication between nodes of optical networks. For integrated communication system integrating above-water and underwater terminals, only free-space optical laser communication can transfer information significant for tactical and strategic applications in the most secure way. This can be achieved in the following way: (1) develop optical laser communication platforms and architecture to transmit and receive from each other in air, and also from underwater platforms which can be friendly submarine, unmanned underwater vehicles (UUV), and autonomous underwater vehicle (AUV); (2) develop the optimum architecture for all-underwater (terminal-to-terminal) terminals communicating with each other, and also through the air-water interface (AWI) capable of communicating with an above-water terminal; and (3) finally optimize the globally distributed FSO communication system mentioned above to the command system for specific strategic application of interest.

In addition, a LADAR (Laser Detection And Ranging) system may be equipped with an underwater optical terminal, which can use a short laser pulse to illuminate and image a target of interest to provide long-range reconnaissance with greater fidelity allowing for the location and image of the detected object and the laser to keep secret. The underwater laser communication system is capable of transmitting different file formats of video, audio, image, and text file and supports rates potentially up to Gbit/s ranges. Recent intense activity in the development of specific new laser sources is suited to specific applications of establishing laser communication with submersible objects. The blue-green laser technology and extremely narrow-band optical filter designs are also useful for space-based detectability and identification/location of submersibles, in addition to laser communications between a satellite and a submerged object. The various absorption and scattering loss components of the satellite-to-underwater target downlink and underwater target-to-satellite uplink using blue-green laser wavelength will also be presented in this section. The reflected returns from a submerged vessel can be picked up by an aircraft, UAV, HAP, or a LEO/small satellite and after advanced signal processing can detect and identify an underwater target of interest. In the discussion, a space-based platform can also include a laser transceiver on a nanosatellite, which can be very lightweight using the recent developments in nanotechnology and ultrasmall, fast optical devices for manipulating transmitting and receiving optical signals in real time.

4.2.3.4   Space-Based Detectability, Identification, and Communication of Underwater Object/Submarine Using Laser Communication

A System and Method to Detect Signatures from an Underwater Object/Submarine

A system and method to detect signatures from an underwater object have been recently discussed in a recent US patent by the author and his colleague [20]. The patent addresses detection of an underwater object from an air-based system, where a broadband acousto-optic signal detection device is associated with the air-based platform. The broadband acousto-optic signal detection device is configured to emit a laser beam toward an underwater object, which can be an underwater vessel or a submarine. The laser beam is reflected back from the underwater object as a return laser beam where the broadband acousto-optic signal detection device detects and receives the return laser beam. One of the main features of this patent is that it addresses the laser beam propagated through the random air-water interface, which is a complex problem. The random air-water interface issue has already been discussed in this chapter earlier. Another interesting aspect of this patent is that the method and the system provide an adaptive optics (AO) technique to mitigate the random air-water interface so that the underwater detection can provide more accurate information about the detectability and identification of submersibles. Adaptive optics technology is a very powerful atmospheric turbulence mitigation technique for FSO communications. The details of AO technique are discussed in this chapter [4]. AO is basically a technology for correcting random optical wave-front distortions in real time. Typically, an adaptive optics system measures the distortions with a wave-front sensor (WFS) and adapts a wave-front corrector to reduce and minimize the phase distortion so that the original signal is reconstructed. The AO technique is designed to measure wave-front errors continuously and correct them automatically. A computer-controlled deformable mirror (DM) can change its shape to correct the wave-front errors, which are measured by WFS that provides correction signal to the mirror. Thus, an adaptive optics system consists of a WFS, DM, computer, controlled hardware, and an advanced software. The reference input signal can generally come from a beacon but not signal laser. There are three types of physical phenomena which can cause random fluctuations in the laser signal between space/aerial-based and underwater platforms: (1) atmospheric turbulence/scattering, (2) random air-water interface, and (3) oceanic turbulence scintillation. The method described in the patent [20] is potentially capable of mitigating all these three types of signal distortions by integrating a laser Doppler vibrometer (LDV) with an adaptive optics (AO) system for sensing and detecting signatures from an underwater object by sensing the acoustic signals originated from the underwater object. A schematic of an LDV system integrated with an AO system for detecting an underwater object (vessel or a submarine) is shown in Fig. 4.8.

Method to Detect Signature from an Underwater Object

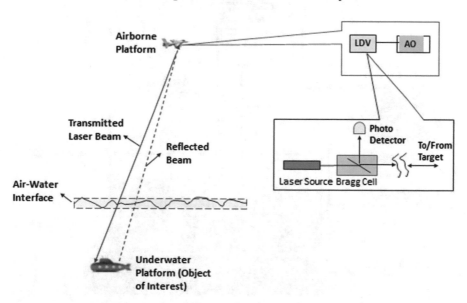

Fig. 4.8 Concept system for detecting signature from an underwater object using laser Doppler vibrometer (LDV) and adapting optics (AO) subsystems

4.2.3.5 Demonstration of a Laboratory Water Tank Experimental Setup for Detecting Underwater Object

Recently, an innovative technique using a laser Doppler vibrometer (LDV) for acousto-optic sensing for detecting acoustic information propagated toward the water surface from a submerged platform has been reported [21]. The authors demonstrated the feasibility of LDV technology by performing a 12-gallon water tank in a laboratory experimental setup as shown in Fig. 4.9. The LDV probes and penetrates the water surface from an aerial platform placed above the water surface. The LDV detected random air-water surface interface vibrations caused by an amplifier to a speaker generating a signal, which is generated from underneath the water surface. The signal generated ranges from 50 Hz to 5 kHz, and the water surface is varied from 1″ to 8″. Information/messages from a submerged platform can be transmitted/sent acoustically to the surface of the water, which then can be detected optically after receiving the information/message using the LDV via the Doppler effect, which thus makes the LDV a highly sensitive optical-acoustic device. The optical signal can be processed to obtain information about the underwater object such as detectability and identification of underwater objects. The laser Doppler

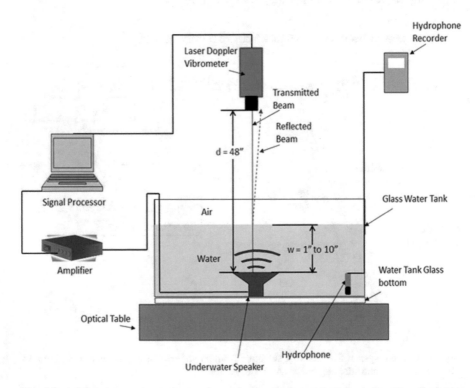

Fig. 4.9 A laboratory water tank experimental setup to demonstrate laser Doppler vibrometer (LDV) technique detecting underwater object from air. (*Reprinted with permission from SPIE Proc. 2021* [21])

vibrometer uses the Doppler effect and optical interferometry for the detection process. The technique is a velocity and displacement measurement technique that uses optical interferometry method, which requires two coherent light beams with their respective light intensities, one transmitted (initial) beam and the other reflected (return) beam, so that they overlap to acquire vibrational waves. The frequency shift of the wave measured by the LDV can be written as $f_D = 2 \cdot v/\lambda$, where v is the object's velocity and λ is the wavelength of the emitted wave which is 632 nm in this experiment. The intensity of the two waves (reference and target beams) can be written as [21]

$$I(t) = (1/2)\epsilon \, c(E_0/2)^2 \left[1 + \cos(2\pi \cdot 2v/\lambda t)\right] \tag{4.25}$$

where ϵ is the dielectric constant of propagation and E_0 is the amplitude function of the wave. The target velocity is related to the Doppler shift of the scattered light by the flowing equation $v(t) = f_D(t)\lambda/(2\cos\alpha)$, where $v(t)$ is the vibration velocity of the target (speaker), $f_D(t)$ is the Doppler shift of the laser light, and α is the angle between the laser beam direction and the vibration velocity vector, which is approximately $0°$ in the experiment.

4.2.4 Example of Laser Communication from a Satellite to Submarine: Impact on Communications Performance

4.2.4.1 Downlink and Uplink Laser Propagation (Air-to-Water-to-Air Channels)

Let the laser transmitter is located at a LEO satellite or an airborne platform sending the optical beam toward an underwater platform. The laser beam is transmitted toward the ocean, which is reflected from the target through the water and onward to the receiver aperture. The received energy per laser pulse at the receiver aperture located at the satellite or aircraft platform is given by [22]

$$E_{\text{rec./pulse}} = E_{\text{trans./pulse}} \div \pi/4(\theta_t)^2 \left[\tau_{\text{atm}}\tau_{\text{cloud}}\tau_{\text{cw}}\right]\left[\tau_{\text{aw}}\tau_{\text{w}}\right]\left[\sigma \cdot \text{Angular scattering}\right]$$
$$\left[\tau_{\text{w}}\tau_{\text{wa}}\tau_{\text{wc}}\right]\left[\tau_{\text{cloud}}\tau_{\text{atm}}\right]\left[1/\pi \, (\theta_r' R)^2\right]\left[A_r\right] \tag{4.26}$$

where the transmittances are denoted by τ_x with x denoting the following suffixes:

x = atm. (atmospheric), cloud (cloud), cw (cloud-to-water), aw (air-to-water), w (water), and
θ_t = beamwidth of transmitted beam
θ_r' = angular scattering bound from the ocean surface
σ = ladar cross section (similar to the *radar cross section* for RF frequency)

A_r = area of receiver aperture

R = distance between laser at the LEO satellite or an airborne platform and entrance into the ocean

4.2.4.2   Submarine-to-Satellite/Aircraft Laser Communication

This is an *uplink* scenario to explore the feasibility of achieving laser communication from underwater vessel/submarine to a satellite (e.g., LEO or a small or nano-satellite) or airborne platform. It is assumed that the submarine is equipped with a laser transmitter pointing toward a satellite or airborne platform first through water, then air-water interface, and finally through atmosphere to reach the optical receiver located on the space/aerial platform. Transmitting from the submarine to the space/airborne platform involves three propagation losses at the operating optical wavelength of interest, i.e., at the laser communication wavelength, which is a blue-green laser (~450–530 nm range). The various losses are due to optical propagation channels in water, air-water interface, and atmosphere and can be summarized as follows:

Water: Absorption and scattering by seawater, beam spreads due to variation in refractive index and suspended biological particle, fading in water

Air-water interface: Random reflection and emergent beams through water, beam spread and beam width increase due to application of Snell's law to random medium

Atmosphere: Atmospheric absorption and scattering due to water and oxygen lines, atmospheric turbulence effects causing beam spread and fading

Oceanic turbulence scintillation and beam spread loss: The power loss due to beam spread from seawater turbulence [23, 24] is given by [22]

$$\text{Power Loss} = -10 \log \left(P_T / P_0 \right)^2$$
$$= -10 \log \left[1 + R^3 \epsilon / \left\{ 3\omega_0^2 \left(1 + 2 / k^2 \omega_0^4 \right) \right\} \right] \tag{4.27}$$

where

P_T = laser transmitter power from a laser transmitter installed in the underwater platform (e.g., submarine)

P_0 = power at the end of the seawater turbulence regime

R = distance between submarine laser transmitter and entrance into ocean

ϵ = turbulence measure of index of refraction variation relative to the radius of turbulence globules, where typical values can be assumed to be about 10^{-10} to 10^{-12} cm^{-1}

ω_0 = radius of Gaussian cross section of laser beam

k = wavenumber = $2\pi/\lambda$

Beam spread loss due to suspended biological particles: When laser beam propagates from the submarine laser transmitter through the seawater, there is also beam spread loss due to suspended biological particles already existing in the water. The power loss can be written as [22]

$$\text{Power Loss} = -10\log\left(P_p / P_0\right)^2 \qquad (4.28)$$

where

P_p = signal power when propagated through suspended bioparticle
P_0 = signal power in the absence of biological particles

and for small-angle scattering length of R_c,

$$\pi P_p^2 = \pi \left[\omega_0^2 \left(1 + R^2 / k^2 \omega_0^4\right) + 8 / 3R / R_c \left(R / kc_p\right)^2 \right]$$
$$R_c = 2n_0^2 / \left\{\left(\pi^{\frac{1}{2}}\right) f \mu^2 k^2 a_p\right\} \qquad (4.29)$$

where n_0 is the refractive index of the water medium, f is the relative volume concentration of the suspended particles, and μ is the refractive index differences between particles and water without particles. The equation for the small-angle scattering length is R_c which is valid for a_p larger than the λ and typically $a_p \geq 10^{-4}$ cm [22].

Some of the typical loss values are reported [22]: For a submarine at 100 ft below the sea surface, the extinction loss of signal power is 16 dB, beam spread due to turbulence suffers a loss of 5 dB, beam spread loss is 33 dB, interface loss is about 3 dB, space loss is ~300 dB, and atmospheric fading loss is 7 dB. Another loss in the presence of clouds in the propagation path may be typically 7 dB or higher.

4.2.4.3 Satellite/Aircraft-to-Submarine Laser Communication

The sources of all the losses from a satellite-to-submarine include the space loss from a satellite to a submarine, loss due to presence of cloud, air-seawater interface, and losses in water including due to biological particles and oceanic turbulence. Typical values of the losses are discussed [22]: the space loss is typically ~300 dB, loss due to cloud is about 4–14 dB, and air-seawater interface loss together with losses in water is about 5–50 dB. Figure 4.10 shows a technology concept of *detectability, identification,* and *duplex laser communications* between satellite/airborne and underwater terminals.

Propagation channel characteristics through the air-water interface for uplink and downlink are *not reciprocal*. Because of multiple scattering effect, an ultrashort laser pulse experiences cloud pulse spreading, which can be typically about 100s of ns to 10s of μs. In addition to temporal spreading, cloud also causes spatial spreading. Finally, seawater also causes further temporal spreading of about 10s of ns.

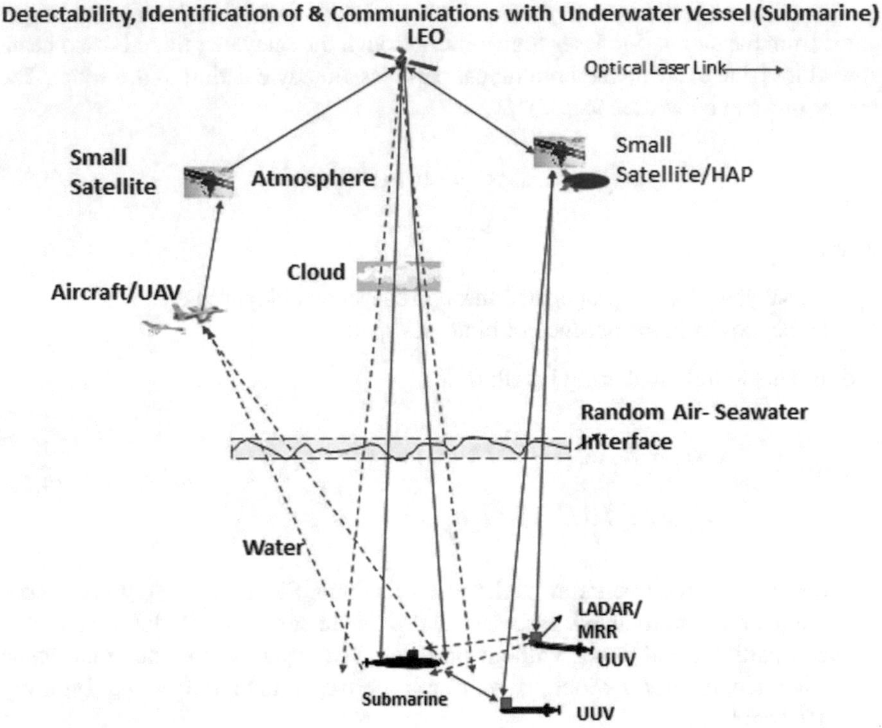

Detectability, Identification of & Communications with Underwater Vessel (Submarine)

Fig. 4.10 Detectability, identification, and duplex laser communications between satellite/airborne and underwater terminals

For downlink laser communication, exponential attenuation of both signal and background occurs at daytime. At night, for uplink only signal is attenuated with depth. In order to improve space/underwater laser communication system performance, a 518 nm Fraunhofer line optical Lyot filter may be used to match the Fraunhofer band. For high-altitude satellite/airborne platform, filters can reduce receiver field of view (FOV) for better bandwidth performance since the signal bandwidth is inversely proportional to the square of FOV: $\Delta\lambda_{\text{sig}} \sim (1/\text{FOV}^2)$.

The aircraft/UAV receiving the laser signal after detecting and identifying an underwater target (e.g., submarine) of interest then relays the signal toward a LEO satellite, which then relays the signal to a command center for further evaluation and action. A number of LEO satellites in a constellation can share and continue to track the signal by relaying the messages.

4.2.4.4   Impact on Free-Space Laser Communication System Performance

For establishing a successful laser communication from a submarine to a LEO satellite or aircraft (airborne platform), it is necessary to evaluate the received signal power from an end-to-end link analysis, which provides the available signal-to-noise ratio (SNR) at the optical receiver at the satellite or the airborne platform. The effective SNR available at the receiver consists of two components: (1) initial transmitting laser power signal (and transmitting angle), and all the losses due to absorption, scattering, etc. due to propagation channels (water, air-water interface, atmosphere), and (2) noise components due to fluctuating signals due to atmospheric turbulence, random air-water interface, and oceanic turbulence. The effective SNR is computed from the ensemble average of fluctuating SNRs, which depend on the probability density functions (PDFs) of laser signal fluctuations for the three media of water, air-water interface, and atmosphere. Note that the physics of optical propagation through different channels are different and therefore the forms of the PDFs are different. Finally, the bit error rate (BER) can be computed from the values of laser-transmitting signal power and the *effective* SNR considering random propagation effects all the way from submarine to satellite propagation.

4.3 Summary and Discussions

This chapter discusses the development of high-speed (10s of Gbit/s or more) optical networks by integrating space (conventional satellite, small and microsatellites, high-altitude platforms (HAPs), UAVs, balloons, airborne), terrestrial, and underwater platforms. The global communication connectivity and Internet access will thus be possible if the airborne satellites and the fixed terrestrial terminals can all be integrated with appropriate optical devices and laser/optical transceivers. Critical issues and challenges of integrating space, terrestrial, and underwater communication links for designing FSO communication system are addressed. Some of the issues discussed are atmospheric turbulence effects on FSO communication system performance, underwater optical communications (uFSO) system performance, and an example of laser communication from a satellite to submarine using blue-green laser wavelength.

References

1. Jennifer C. Ricklin, Stephen M. Hammel, Frank D. Eaton, and Svetlana L. Lachinova, "Atmospheric channel effects on free-space laser communication", in "Free-Space Laser Communications: Principles and Advances". Springer, New York (2008).
2. Isaac I. Kim and Eric Korevaar, "Availability of Free Space Optics (FSO) and hybrid FSO/RF systems, Proc. SPIE, *Optical Wireless Communications IV*, Eric Korevaar (Editor), 27 November 2002.
3. Arun K. Majumdar, *Optical Wireless Communications for Broadband Global Internet Connectivity: Fundamentals and Potential Applications*, Amsterdam, Netherland Elsevier (2019).
4. Arun K. Majumdar, *Advanced Free Space Optics (FSO): A Systems Approach*, Springer Science+Business Media, New York 2015.

5. Larry C. Andrews and Ronald L. Phillips, *Laser Beam Propagation through Random Media*, Second Edition, SPIE, Bellingham, Washington (2005).
6. H. Kaushal, V. K. Jain, S. Kar, Overview of Wireless Communication Systems, Chapter, January 2017.
7. Arun K. Majumdar, Unpublished Report, January 2021.
8. A. K. Majumdar, T. M. Shay, Wide Field-of-View Amplified Fiber-retro for secure High Data Rate Communications and Remote Data Transfer, U.S. Patent No. 8, 301,032 B2, October 30, 2012.
9. Mark Alan Chancey, "Short Range Underwater Optical Communication Links," MS Thesis, North Carolina State University Electrical Engineering, Raleigh, NC 2005.
10. Korotkova O., Farwell N., and Shchepakina E., "Light scintillation in oceanic turbulence", Waves in Random and Complex Media, Vol. 22, No. 2, pp. 260–26, (2012).
11. V. V. Nikishov and V. I. Nikishov, "Spectrum of turbulent fluctuations of the sea-water refraction index," International Journal of Fluid Mechanics Research, vol. 27, no. 1, pp. 82–98, 2000.
12. N. Farwell, "Optical beam propagation in oceanic turbulence," PhD dissertation, 2014.
13. Jamali M.V., Khorramshahi P., Tashakori A., Chizari A., Shahsavari S., AbdollahRamezani S., Fazelian M., Bahrani S., and Salehi J.A., (2016) "Statistical distribution of intensity fluctuations for underwater wireless optical channels in the presence of air bubbles", Iran Workshop on Communication and Information Theory IWCIT 2016, 3–4 May 2016.
14. Arun K. Majumdar, John Siegenthaler, Phillip Land, "Analysis of optical communications through the random air-water interface: feasibility for underwater communications," Proc. SPIE 8517, Laser Communication and Propagation through the Atmosphere and Oceans, 85170T (24 October 2012); doi: https://doi.org/10.1117/12.928999.
15. S. Jaruwatanadilok, "Underwater wireless optical communication channel modeling and performance evaluation using vector radiative transfer theory," Selected Areas in Communications, IEEE Journal on, vol. 26, pp. 1620–1627, December 2008.
16. B. Cochenour, L. Mullen, and A. Laux, "Spatial and temporal dispersion in high bandwidth underwater laser communication links," in Proc. IEEE Military Communications Conf., pp. 1–7, 2008.
17. L. Mullen, B. Cochenour, W. Rabinovich, R. Mahon, and J. Muth, "Backscatter suppression for underwater modulating retroreflector links using polarization discrimination," Appl. Opt. 482, 328–337, 2009.
18. Shlomi Arnon, "Underwater optical wireless communication network," Optical Engineering 49(1), 015001 (January 2010).
19. S. Arnon and D. Kedar, "Non-line-of-sight underwater optical wireless communication network," J. Opt. Soc. Am. A 263, 530–539 2009.
20. Phillip P. Land and Arun K. Majumdar, "System and Method to Detect Signatures from a Underwater Object," United State Patent, Patent No.: US 10,228,277 B1, Mar.12,2019.
21. Phillip Land, James Roeder, Dennis Robinson and Arun K. Majumdar Radar Sensor Technology XIX; and Active and Passive Signatures VI, Proc. of SPIE Vol. 9461, 94611H, 2015.
22. David G. Aviv, *Laser Space Communications*, ARTECH HOUSE, Inc. Norwood, MA 02062, 2006.
23. Yura, H. T. "Propagation of Finite Cross-Section Laser Beams in Sea Water," Applied Optics, Vol. 12, No. 1, January 1973.
24. Yura, H. T. "Mutual Coherence of a Finite Cross Section Optical Beam Propagating in a Turbulent Medium," Applied Optics, Vol. 11, No. 6, June 1972.
25. XIAOYAN LIU, SUYU YI, XIAOLIN ZHOU, ZHILAI FANG, ZHI-JUN QIU, LAIGUI HU, CHUNXIAO CONG, LIRONG ZHENG, RAN LIU, AND PENGFEI TIAN, "34.5 m underwater optical wireless communication with 2.70 Gbps data rate based on a green laser diode with NRZ-OOK modulation," Optics Express, Vol. 25, No. 22 | 30 Oct 2017 | OPTICS EXPRESS 27937.
26. H. R. Anderson, *Fixed Broadband Wireless System Design* (Wiley & Sons, West Sussex, England, 2003).

Chapter 5
Technology Developments, Research Challenges, and Advances for FSO Communication for Space/Aerial/Terrestrial/Underwater (SATU) Links

5.1 Introduction

Free-space optical communication technology developments and advances have occurred over the past decade leading to wideband optical networks to meet the continuous increasing demands for high-data-rate information exchange. The wireless communication applications have been driven by the ever-increasing number of wireless broadband Internet, mobile phones, smart devices, and social web such as Facebook, WhatsApp, and video gaming and sharing. The recently developed 5G and future 6G wireless networks are addressing the increased data traffic due to widespread use of smart devices like cell phones *anytime, anywhere*. Free-space optical (FSO) wireless communications technology is the only potential solution to meet the user's ever-increasing demand for bandwidth. In the last few chapters of this book, basics of wireless communications relevant to satellite links, optical propagation relevant to integrated space/aerial/terrestrial/underwater links, fundamentals and applications, and optical laser links in space-based systems for global communications architecture have all been addressed and discussed. Need for FSO communication link is also evident because of the most recent space exploration of NASA: NASA's rocket was launched on 30 July 2020, and the rover successfully landed on Mars on 18 February 2021 and sent images to the Earth. In fact, NASA is building a new antenna to add to its deep space network (DSN) for transmitting messages to Mars astronauts, and lasers can boost the data rate from the Red Planet, almost 170 million miles away, ten times more than radio, and will provide a platform for optical communications to encourage other space explorers to experiment with lasers on future missions [1]. Interest in manned missions to the planets would require robust bidirectional high-rate links to support the human infrastructure, telemetry, and Web-based connectivity. However, in spite of the substantial benefits of FSO, all these new technology developments and future developments face significant challenges to the system designers. One of the many challenges specially for long-range FSO communication is due to uncontrollable random atmospheric

© Springer Nature Switzerland AG 2022
A. K. Majumdar, *Laser Communication with Constellation Satellites, UAVs, HAPs and Balloons*, https://doi.org/10.1007/978-3-031-03972-0_5

turbulence channel effects on communication links such as stochastic intensity fluctuations in a received optical signal affecting communication performance parameters like signal-to-noise ratio (SNR), bit error rate (BER), probability of fade, and precise pointing and tracking of narrow transmit beams.

This chapter discusses current advanced research relevant to "all-optical" communications and connectivity and future directions to make the concept of constellation communication platforms successful, which include (1) fast luminescent detector with omnidirectional sensitivity for FSO, (2) nonlinear silicon photonic digital processing, (3) high-speed photonic components using graphene materials for optical switching and optical routing, (4) photonic devices for 100 Gbit/s optical links, (5) on-chip twisted emitter for FSO communication system (ultra-broadband multiplexed orbital angular momentum (OAM) emitter for OAM communications with a data rate of 1.2 Tbit/s), and (6) quantum Internet: hybrid quantum information network/optical chip-integrated smartphones and similar photonic areas. All of these recent researches will contribute to the robust constellation communication in adverse situations.

This chapter presents various technology developments, research challenges, and advances in FSO communication system as well as discusses research challenges to be addressed and solved to overcome the limitations of existing systems and emerging new technologies. The recent applications related to FSO wireless communication systems also include space (aerial)-based and underwater platforms, 5G/6G networks, smart cars, robots, optical links for aiding health-related services, and artificial intelligence (AI)-based machine-to-machine (M2M) devices for adaptive optics (AO). Some of the challenges addressed in this chapter include the following categories:

- Systems and applications (in-orbit, on-ground demonstrations and results, communication systems and scenario, laser communication technology for next-generation applications, ranging technology with optical communications)
- Devices, components, and subsystems (optical devices for space-based applications, optoelectronic subsystems and components, laboratory demonstration hardware)
- Basic link technologies (atmospheric propagation, transmission effects and compensation techniques, site diversity techniques, modulation formats for space-based systems)
- Advanced technologies (advances in quantum communications, novel optical devices and systems for lasercom, new mm-wave/THz communication devices and systems)
- Mobile ad hoc and smart systems using FSO communication
- Innovations in LED wireless communication systems: signify research and signify Li-Fi systems

5.2 FSO Communication Systems and Scenarios: Laser Communication Technology for Next-Generation Applications and Integration with 5G/6G Systems

With increasing number of various application scenarios of FSO wireless communication, there are research challenges to improve system designs for improving communication performances of in-orbit, on-ground, and underwater platforms. Laser communication technology with optical communications is advancing for next-generation applications. Efficient optical networks with low latency are required for integrated space/aerial/underwater laser/optical links.

- The goal of an FSO communication system is to provide high-speed, improved-capacity, cost-effective, and easily deployable wireless networks. A next-generation FSO communication system utilizing seamless connection of free-space and optical fiber links is reported [2]. In their research, the authors described a compact optical antenna utilizing a miniature fine positioning mirror (FPM) for high-speed beam control and steering, which can mitigate power fluctuations at the fiber coupling port due to beam angle-of-arrival (AOA) fluctuations. The system shows a potential data rate from 2.5 to 10 Gbit/s. The research described the transceiver incorporating a FPM for high-speed beam tracking at 1 kHz and control function. Using the wavelength division multiplexing (WDM) technology, the system bandwidth was increased.
- The integration of 5G wireless and optical technology is one of the advanced communication networks which require heterogeneous emerging technologies for future applications. Recent developments of advanced optical networking to provide 5G transport networks and their applications in connecting a large number of devices in future smart city infrastructure are discussed [3].

5.2.1 Key Challenges for 5G/6G Communications and IoT Solutions

Some of the challenges for "beyond 5G or 6G" using laser/optical links are discussed in this section. The solutions from these challenges will be helpful to both urban and rural populations to connect the connected. More recent technologies for 5G [4, 5] and beyond and the latest advances related to backhauling with UAVs, balloons, and HAPs as well as the latest trends and breakthroughs in satellite-based FSO communication connectivity are discussed. Space/aerial-based platforms can be considered as general backhaul terrestrial including multihop or local optical network that can be a part of fronthaul in conjunction with the Wi-Fi (Li-Fi for optical) hotspot (mesh) networks for direct access. One of the challenges for terrestrial FSO is that communication performance is severely limited affected by weather conditions such as fog and alignment errors. Optical propagation through fog specially through dense fog is determined by multiple scattering process, and effective mitigation through fog is still not fully solved. The concept of radio over FSO (RoFSO) system has been reported, in which signals over a fiber backbone are transferred to FSO and the RF signals are modulated over optical carrier. However, the system still is affected by rainfall and scintillation effects. FSO backhaul communications using satellites, HAPs, balloons, and UAVs can be achieved at altitudes where they are less sensitive to weather conditions such as fog or cloud blockage. Although the platforms are very expensive now, they can eventually be significantly less expensive when the platforms will be powered by solar radiation. For very-high-throughput satellite systems, optical feeder links based on DWDM to increase the throughput between ground stations and GEO satellites are discussed, which can significantly reduce the number of needed stations with Ku (12–18 GHz) and Ka (26–40 GHz) bands. Satellite-equipped laser/optical transceivers can be effective to accomplish these goals of achieving high-data-rate communications *anywhere, anytime* to almost any urban, rural, and remote locations. Some of the typical *challenges* for FSO communications between space/aerial and optical ground stations include the atmospheric effects (absorption, scattering, turbulence, presence of fog and cloud blockage), beam divergence loss (due to diffraction close to receiver's aperture), and background noise from the Sun (either direct or scattered). For inter-satellite communications, the challenges are mainly due to pointing, acquisition, and tracking (PAT), Doppler shift due to the relative movements of the satellites, and satellite vibration. All these challenges can be addressed and solved with today's optical technology and devices developed so that FSO communications can be accomplished for GEO-LEO or HAP/UAV/balloons-to-LEO or GEO for high-data-rate, high-capacity communications with small SWP (size, weight, and power).

What exactly are the issues for OWC as a promising solution for data traffic applications?

5G is an example where OWC can provide a solution in the ultradense heterogeneous networks for which mobile data volume per area will be 1000 times and the number of connected wireless devices will be 100 times higher compared to those existing networks [6]. The future networks should therefore need to support not only high user data rates but also low power consumption and negligible end-to-end delays (latency). As discussed earlier, both backhaul and fronthaul connectivities capable of offering high-data-rate and high-capacity communications are absolutely needed for ultrafast access networks. Also, the Internet of Things (IoT) currently being developed will require even more higher Internet connectivity demand because more and more physical devices will then be connected. Some of the OWC technologies needed to support the huge heterostructure network demand include visible light communication (VLC), light fidelity (Li-Fi), optical camera communication (OCC), efficient FSO communication uplinks and downlinks, and compact light detection and ranging (LiDAR).

Some of the potential challenges for incorporating FSO wireless communications that need to be addressed are the following [7]:

- *Handover* between two optical and RF networks since the physical and data link layers are different for the two networks. Consequently, there is the challenge for mobility support for the integrated hybrid network system.
- *Interference* between VLC and Li-Fi network cells.
- *Uplink* communication for VLC and Li-Fi systems is harder than the downlink communication because of the spread of LED diffused lights and the optical geometries are different and not reciprocal for uplink and downlink configurations.
- Existing *low-frame-rate* cameras cannot provide the required high-data-rate demands for 5G/6G and IoT networks.
- *FSO backhaul* system needs to be improved for achieving efficient communication capacity because of the high-data-rate exchange of enormous amount of data.
- *Artificial intelligence* (AI)-based *machine-to-machine* (M2M) networks need to be incorporated for 6G communication network.

FSO wireless communication system has the best potential for handling the above challenges. Portable and extremely small and compact optical devices such as very fast optical switching, signal processing with optical chips, extremely small LED and laser diodes offering various optical wavelengths for mitigating atmospheric effects, omnidirectional optical antennas, and extremely highly sensitive optical detectors are just some of the recent developments to meet all the challenges discussed above. Recently, the vision of future 6G wireless communication and its network architecture has been discussed [8]. The authors addressed some of the fundamental enabling technologies for 6G as follows: (1) *artificial intelligence (AI)*—the technology will play a vital role in M2M, machine-to-human, and human-to-machine communications where a significant reduction of latency can be achieved; (2) *terahertz communications*—this technology will enhance the 6G potentials and support wireless cognition, sensing, imaging, communication, and positioning; (3) improved *FSO fronthaul and backhaul network* will improve connectivity and will be able to improve limitations by atmospheric effects; (4) *MIMO* technology to better spectral and energy efficiency and for higher data rate; (5) *3D networking* by integrating satellite, UAV/HAP, terrestrial, and underwater networks for global coverage; and (6) highly secured communications such as *quantum communications*. These technologies are required to move from 5G to future 6G and Internet of Everything (IoE) communications. Figure 5.1 shows an example of optical wireless communication network for the 5G/6G and Internet of Things (IoT) platforms, which include and integrate space, aerial, terrestrial, and underwater communications. The interested readers can read the excellent survey paper [8], which discusses the above technologies in details.

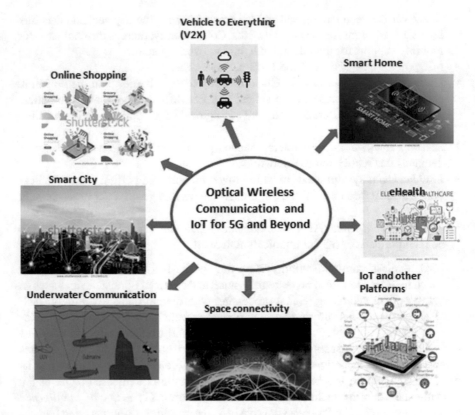

Fig. 5.1 Optical wireless communication networks for the 5G/6G and IoT platforms integrating space, aerial, terrestrial, and underwater communications

5.3 A New Technique of Deep Neural Network (DNN) for Sensing and Verifying Atmospheric Turbulence Parameters for FSO Communication Channel

Recent research on a machine learning-inspired atmospheric turbulence sensing is reported [9] using deep neural network (DNN)-based processing of short-exposure laser intensity scintillation patterns. The machine learning approach provided the atmospheric sensing with good accuracy and high temporal resolution. For predicting FSO communication system performance accurately, it is important to analyze the atmospheric turbulence dynamics with sufficiently high temporal resolution. Existing electro-optical C_n^2 sensors have a fundamental drawback since the sensing method depends on the conventional Kolmogorov's theory with Taylor's frozen turbulence hypothesis, which are not strictly correct for near-ground and long-range slant-path scenarios. The authors described the machine learning experiments utilizing datasets of a large number (up to 1.2×10^5) of data instances consisting of C_n^2

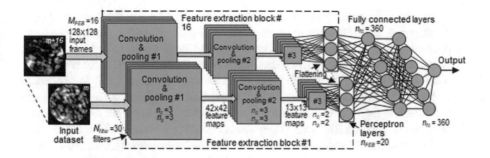

Fig. 5.2 The topology of the C_n^2 net architecture model used for C_n^2 prediction based on DNN processing of short-exposure intensity scintillation images. (*Reprinted with permission from MDPI Support-Open Access, 2021* [9])

values and laser beam intensity scintillation images. Their DNN architecture and implementation are summarized below:

- DNN is designed for spatial feature extraction from image streams compromised of a sequence (stack) of convolution and pooling image processing layers. A 2D array of data (feature map) processed by the convolutional layer enters a corresponding pooling layer.
- The authors utilized the general building blocks in the DNN topology, designed and optimized for C_n^2 prediction based on DNN-based processing of short-exposure intensity scintillation images.
- Their C_n^2 net architecture, training, validation, and performance evaluation were discussed in detail in their report.
- DNN-inspired atmospheric turbulence characterization and machine learning-based turbulence profiling and turbulence model evaluation via machine learning are explained in details.

In order to reduce the dimension (subsample) of feature maps, the major goal of pooling layers, for example, the so-called max-pooling layers, only the max value of the pooling kernel of size $n_p \times n_p$ propagates to the next computational layer which reduces the dimension of the feature map by a factor of n_p^2. Figure 5.2 shows the topology of the C_n^2 net architecture model discussed in their research paper [9] used for C_n^2 prediction based on DNN processing of short-exposure intensity scintillation images. The C_n^2 net model has $M_{FEB} = 16$ feature extraction blocks (FEBs) with identical topology and trainable weights where each FEB is composed of three convolutional and max-pooling layers that succeed each other and a single 1D input layer (perception layer) composed of $n_{FEB} = 20$ neurons. At the output, vectors of all FEBs are merged into a single feature vector that inputs two fully connected layers comprised of $n_{fc} = n_{FEB} \times M_{FEB} = 360$ neurons as explained in [9].

The research study [9] clearly shows that an optical system compromised of a remotely located laser beacon and an optical receiver acquiring short-exposure intensity scintillation patterns digitally processed with DNN can be used for in situ atmospheric turbulence characterization specially for long-range and slant-path FSO communication system even in strong turbulence regime. This scenario is applicable to satellite/aerial/UAV to optical ground station links in order to evaluate the communication system performance such as available SNR, BER, or probability of fades. The technology scheme described in their research work is especially important for prediction and in situ adjustment of electro-optical system parameters based on atmospheric sensing information. As an example, the prediction of coming deep signal fading in a free-space laser communication system can be used to minimize losses in the throughput data to improve FSO communication performance. This is also applicable to an adaptive optics (AO)-based free-space laser communication system where in situ prediction of turbulence conditions occurring within the timescale of a few seconds can be used to adjust the parameters of turbulence-induced distorted wave front and control system in almost real time. The machine learning approach for characterizing atmospheric turbulence offers a wide range of capabilities for *real-time* fusion of the data flows coming from various optical sensors. This can be important in all the scenarios where satellite/aerial/UAV/HAP platforms are moving as well as a constant dynamic evaluation of atmosphere turbulence between these platforms and fixed optical ground stations located in different places.

5.4   Neural Network Technology for Improving FSO Communication in the Presence of Atmospheric Turbulence and Distortions

Advanced neural network technology has been recently developed [22] by combining generative neural network (GNN) and convolutional neural network (CNN) system [40] in order to improve received signals severely deteriorated by atmospheric turbulence and other noise in a FSO communication system. The proposed technology which has been demonstrated by both simulated and experimental communications settings involves (1) correction for signal distortions and reduction of detector noise with GNN system to reproduce almost exactly the desired mode profiles at the communication receiver and (2) improving the classification accuracy after demodulating the generated modes using a CNN system that is pretrained with simulated optical profiles.

The method is based on using orbital angular momentum (OAM), which allows applying large alphabets in optical communication schemes by generating and transmitting various superpositions of OAM states and thus significantly increasing the alphabet sizes of the system. The effective demodulation of the received signal involves comparing and then identifying which OAM superposition is sent and received. In an FSO communication system due to the presence of turbulence-induced random distortions, the classification efficiency is restricted. The researchers developed the generative network by demodulating the signal effectively using a CNN classifier trained with only undistorted optical modes with added dark noise [22]. A significant enhancement using the GNN was also shown by calculating the cross talk between the noisy and distorted modes at the receiver. The method thus requires no turbulence corrections by adaptive optics (AO), and the network architecture can be pretrained and portable with less complexity than AO system. Figure 5.3 shows the network architecture and reconstruction of various orbital angular momentum (OAM) modes consisting of a generative neural network (GNN) and a convolutional neural network (CNN). The range of turbulence strength for the demonstration was $C_n^2 = 22.4 \times 10^{-11}$ m$^{-2/3}$ to $C_n^2 = 80 \times 10^{-11}$ m$^{-2/3}$, and the effectiveness of the method was shown by comparing the images of various noisy experimental optical orbital angular momentum (OAM) modes with increasing attenuation ratio in the range of −0.18 to −2.43 dB.

5.5   Highly Sensitive Optical Receivers for FSO Communication

A general scheme of FSO communication with I–Q modulator and demodulator is shown in Fig. 5.4. Free-space optical wireless communication demands the most sensitive receivers possible for maximum long-range reach while also requiring high-data-rate operations. Researchers have recently reported the development of a receiver using an almost noiseless optical preamplifier for the application of FSO communications. Sensitive optical receivers for space have been discussed recently [11]. The laser/optical transmission system relies on an optical amplifier in the receiver which does not add any excess noise, in contrast to all other existing optical phase-sensitive amplifiers. The new concept developed at Chalmers University of Technology in Sweden demonstrates [10] a receiver sensitivity of just one photon per information bit at a data rate of 10.5 Gbit/s. This concept will be extremely

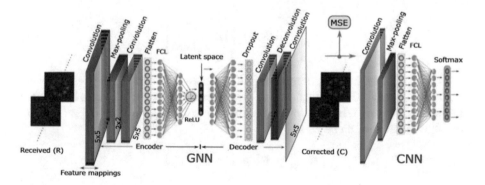

Fig. 5.3 Network architecture of generative machine learning consisting of a generative neural network (GNN) and a convolutional neural network (CNN). (*Reprinted with permission from Communication Physics 2021* [22])

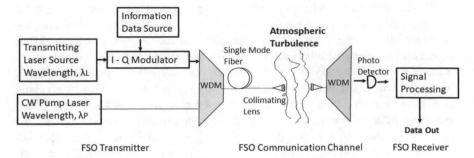

Fig. 5.4 FSO communication with *I–Q* modulator and matching demodulator

important to both short- and long-range optical communications to achieve the lowest bit error rate (BER) at the desired high data rate in the presence of other sources of noises including due to atmospheric turbulence and scattering. The system developed by the researchers can easily adopt a simple modulation format of quadrature phase-shift keying (QPSK), forward error correction (FEC) coding, and standard digital signal processing to recover data. A summary of the concept is discussed below:

The theoretically achievable minimum sensitivity of a detector can be found by analyzing the Shannon's channel capacity. In practice, no channel is noiseless, and the capacity C cannot be increased infinitely and is limited by Shannon's channel capacity for noisy channel and still further limited by receiver constraints and noise sources:

$$C = B \log_2 \left(1 + S / N \right)$$

where *S/N* is the signal-to-noise ratio.

The maximum information rate under error-free data transmission for a preamplified dual-quadrature or phase-diverse coherent homodyne receiver is given by [10]

$$C_{\text{preamp}} = B \log_2 \left(1 + \frac{2S}{F_N h \nu B} \right) \tag{5.1}$$

where C_{preamp} is the capacity of a preamplified receiver, S is the signal power, h is the Planck's constant, ν is the frequency of the optical carrier wave, and B is the inverse of the symbol period. In terms of the number of photons per transmitted symbol, n_s, the above equation can be written as

$$C_{\text{preamp}} = B \log_2 \left(1 + 2 n_s / F_N \right)$$

where F_N is the noise figure of the preamplifier. The term $2n_s/F_N$ is therefore equivalent to the signal-to-noise ratio (S/N) mentioned earlier. For an erbium-doped fiber amplifier (EDFA), the possible noise figure is 3 dB; however, the phase-sensitive

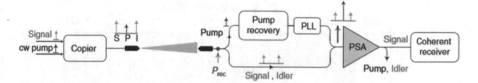

Fig. 5.5 FSO communication system with a sensitive receiver using phase-sensitive amplifier (PSA). (*Reprinted with permission from Light: Science & Applications 2021* [10])

optical amplifier (PSA) can have a theoretical noise figure of 0 dB, which can thus amplify the signal without *excess noise*. The experimental technique reported in [10] used the one-pump nondegenerate signal/idler configuration to amplify the quadrature of an optical wave, which requires two input waves at different wavelengths. The overall quantum-limited NF of the PSA was still 0 dB. The capacity of the PSA is given by

$$C_{PSA} = \frac{B}{2}\log_2\left(1 + 4n_s\right) \tag{5.2}$$

In order to determine the ultimate sensitivity, the researchers calculated the ratio of the number of photons per symbol n_s to the number of bits per symbol resulting in the best possible sensitivity for the PSA to be 0.35 PPB. In the experiment, the researchers reported that PSA can reach a black box sensitivity of 1 PPB. In the reported FSO communication experimental setup, the data was modulated using quadrature phase-shift keying (QPSK) modulation scheme to transmit a net information at 10.52 Gbit/s. A conjugate idler wave was generated by combining the signal with a CW pump in the copier stage. The three waves, the signal (λ_s), idler (λ_I), and pump (λ_P), were launched to the fiber collimator to send to the free-space channel. In the experiment reported, the results were obtained from a laboratory setup using 1 m link. The received power, P_{rec}, after the fiber collimator contained the total power of the signal, idler, and pump. The pump power was then separated from the signal and idler with a wavelength division multiplexer (WDM). A highly nonlinear fiber (HNLF) combined the signal, idler, and recovered pump for phase-sensitive amplification of the signal. After filtering the PSA output signal, the information signal was detected using a coherent receiver and processed with DSP to recover the information. A general diagram of a potential FSO communication system is shown in Fig. 5.5. The reported [10] results of achievable BER as a function of received photons/symbol in dB are shown in Fig. 5.6, which indicate that with a PSA preamplifier, a transmission with a received power of 1 photon/symbol or 1 photon/information bit (PPB) can be achieved with a bit error rate (BER) below 10^{-6}, which is the best sensitivity reported to date. The results are extremely encouraging to design FSO communication system for long-range operating in the presence of atmospheric turbulence with forward error correction (FEC).

This new development of potential long-range FSO communication systems with sensitivity approaching 1 single photon/bit will be extremely useful in

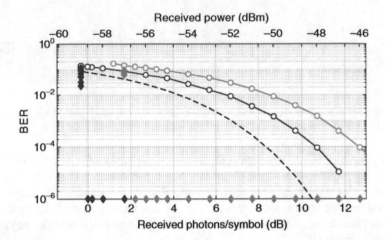

Fig. 5.6 Results for achievable BER as a function of received photons/symbol in dB. (*Reprinted with permission from Light: Science & Applications 2021* [10])

designing FSO communication systems to operate under atmospheric turbulence conditions and various potential application scenarios (ground-to-space/aerial-based platforms, 5G/6G) where high data rate is required using average power-limited photon-starved optical links. This concept technology uses only standard single coherent receiver, but *multi-aperture coherent combination* approach is also being considered [12]. Finally, this highly sensitive receiver technology will help for substantially increasing the reach and high data information rate for future high-speed links, which will be directly applicable to inter-satellite communication or between two satellites in a constellation of satellites and sending back the information from a satellite to optical ground station through atmospheric turbulence.

5.6 Highly Sensitive FSO Communications Using Chip-Scale LED Transmitters and Single-Photo Receivers for Small Satellite Platforms

Recently, FSO communication link implemented with a CMOS-controlled micro-LED transmitter and a single-photon avalanche diode (SPAD) array receiver are reported [13], which are well suited for small platforms where small size, weight, and power (SWaP) are critical. For FSO communication applications, compact integrated devices with a digital interface can be constructed using micro-LEDs fabricated in high-density format and bump bonded to complementary CMOS control electronics. The micro-LED chip integrated with CMOS control electronics allows modulation, control, and power to be supplied by digital signals applied directly to the device. The micro-LED system could be a single 10×10 cm^2 printed circuit board (PCB) or possibly can be even mounted on the satellite chassis. Single-photon receivers can provide exceptionally high sensitivity closer to the standard quantum limit (SQL). In order to add a mitigating code such as forward error correction (FEC), typically a bit error rate (BER) of about 2×10^{-3} and 1×10^{-2} (uncoded) value is needed so that FEC code can be activated to achieve a BER of 10^{-9} or smaller. The SQL of a communication system using return-to-zero (RZ) on-off keying (OOK) modulation method is written as

$$\text{SQL} = -\frac{hcR}{2\lambda}\log(\text{BER}) \tag{5.3}$$

where h is the Planck's constant, c is the speed of light, λ is the optical carrier wavelength, R is the bit rate. Sensitivity limit at a BER of 2×10^{-3} for given data rates using RZ-OOK modulation and the result from a 750 m link of reaching a lower BER of 1×10^{-5} are discussed [13]. Data rates up to 100 Mbit/s are demonstrated with a sensitivity of -55.2 dBm with a difference of 13.42 dBm from the SQL. The sensitivity limit at a BER of 2×10^{-3} for given data rates using RZ-OOK encoding is reported where a separation in dB from the SQL is also depicted. Simulated achievable data rates for a given range, based on the intensity data, are also reported. The receiver consists of 64×64 SPADs on a 21 μm pitch and operated at room temperature. The chip is 2.6×2.8 mm^2 and is packaged to interface with a PCB. The control signals and power to the chip were provided by a FPGA. The individual pixels have a photon detection probability of 26% at 450 nm.

The researchers demonstrated optical communication link performance for short ranges and data rates in the ranges of 100 Mbit/s (laboratory) and about 20 Mbit/s (terrestrial). The results from the research clearly provide a pathway for designing very compact highly sensitive transceivers for improving FSO communication performance for deployment of recent small and CubeSats for space/aerial platforms. Highly sensitive receivers which are very compact together with optimized transmission schemes will enable improved FSO data transmission requiring very low levels of received power. The reported work discusses chip-scale transmitter and receiver elements for the inter-satellite links to develop transceiver systems occupying a fraction of 1U $((10 \text{ cm})^3$, 1 kg "1U" units) achieving up to 52 dBi directional gain [13]. As shown in Fig. 5.7, the data rates of 100 Mbit/s with sensitivity of -55.2 dBm in the laboratory and a 20 Mbit/s for a terrestrial range of 750 m with combined power consumption of the system of less than 5.5 W were achieved. Another advantage of using micro-LED sources in array format is that because of multiple emission beams, angular alignment of small satellites would be much easier to achieve the required pointing accuracy without increasing SWaP footprint. Future chip-scale transmitter technologies may include vertical cavity surface-emitting lasers (VCSELs), which can also be fabricated in array format to offer narrow-emission linewidth (helpful in rejecting background noise) and have the capability to offer high-modulation bandwidths. For highly sensitive receiver point of view, blue-emitting VCSELs are still not technologically matured for the application of inter-satellite links (ISLs). Near-infrared VCSELs are also not well matched to the peak sensitivity of silicon SPADs. The concept technology can also be applied for

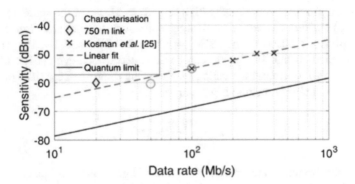

Fig. 5.7 Sensitivity limit shows the experimental results and difference in dB from the quantum limit. (*Reprinted with permission from The Optical Society of America, Optics Express, 2021* [13])

achieving *underwater free-space communication* (uFSO) using the blue-green light around 450 nm, which will be suitable for a short range in the seawater requiring low data rate.

5.7   Terabit/s-Level Optical Data Transmission for FSO Communication: Laser Transmitter Development Technologies

5.7.1   *1.72 Tbit/s Optical Data Transmission Demonstration*

Researchers at the German Aerospace Center DLR reported [14, 15] the world record in free-space laser/optical communications for satellite applications at 1.72 Tbit/s across a distance of 10.45 km. This clearly paved the way for developing future global broadband Internet connectivity in remote areas of the world using *all-optical* technology. The successful transmission at 1.72 Tbit/s across a distance of 3 km in free space was reported by the DLR Institute of Communications and Navigation. Another successful test between ground and mountain for a propagation path length of 10.45 km, where atmospheric effects like turbulence and scattering media affect the laser communications performance, was also performed. The atmospheric test results were very useful for the DLR scientists and engineers to develop a new free-space optic transmission system that can be a stable optical link for long-range laser communication with the satellite. Recently, the researchers demonstrated [16] 1.72 Tbit/s output using optical dense wavelength division multiplexing (DWDM) method. The paper discussed modulating 40 DWDM channels with uncoded 43.01824 Gbit/s rate per channel. Forty carriers spaced at 100 GHz grid emitting in the optical C band were modulated using on-off keying (OOK) with a known pseudorandom bit sequence (PBRS) at 43.01824 Gbit/s, de-correlated and amplified to a total transmit power of 32 dBm. The transmitting beam was a

Gaussian beam with a divergence of 100 μrad. This proof-of-concept demonstration of 1.72 Tbit/s transmission through turbulent atmosphere is similar to a ground-to-GEO satellite optical uplink.

5.7.2   1.28 Tbit/s WDM Transmission for Free-Space Optical Communications

A compact free-space optical terminal (FSOT) was demonstrated [17] to support a high transmission capacity of 1.28 Tbit/s using 32 CW DFB lasers, WDM with intensity modulating (nonreturn to zero) each channel at 40 Gbit/s each, and then amplified by EDFA to transmit over an FSO link. The CW DFB lasers were capable to transmit 100 GHz spacings (from 1535.7 to 1560.5 nm) and were aligned in polarization, combined by an arrayed waveguide grating (AWG), and used a LiNbO$_3$ intensity modulator. The demonstration is important to develop terabit-level data transmission for improved stable (6 h with no errors) FSO communications.

5.7.3   FSO Communication Using All-Optical Retro-modulation at Terabit/s Data Rates

A new concept in modulated retroreflector (MRR) technology with fiber-based amplified retro-modulator is already reported [18]. An all-optical retro-modulation device and system architectures for FSO wireless communication at terabit/s data rate have been presented [19], which is capable of enabling multidirectional FSO communication. Future FSO wireless communication technology can be tremendously improved using *all-optical* networks, which can solve electronic bottlenecks allowing further network transparency and achieve terabit/s-level data rate. One of the important challenges in accomplishing duplex or bidirectional communication links is the requirement of line-of-sight (LOS) alignment with sub-microradian accuracy specially needed for multidirectional asymmetrical networks. A typical scenario can be to establish high-data-rate links from one transceiver to multiple distributed transceivers with all-optical technology. The researchers discussed the implementation of such system architecture utilizing a passive uplink via retro-modulation where the retro-modulator is activated by the received optical power from an active downlink. This concept technology can eliminate the pointing-acquisition-tracking while supporting communication with a number of transceivers within the field of view (FOV) of the retroreflector. The researchers describe an all-optical implementation of passive uplinks, which integrates an all-optical modulator. A spherical (cat's eye) retroreflector was designed for realizing a wide FOV. An all-optical modulation scheme at the passive transceiver was employed, which was implemented through cross-absorption modulation where a control (pump) beam was used to modulate the absorption seen by a coincident signal (probe) beam. A low-cost thin film of cupric oxide (CuO) nanocrystalline material was implemented with a spherical retroreflector to design the all-optical retro-modulation device for passive uplinks.

The signal at the active transceiver for a transmitting collimated beam after modulation and retroreflected back is given by [18, 19]

$$P_d = \eta P_{tx} M \frac{A_s}{\Omega_{tx} R^2} \cdot \frac{A_r}{\Omega_{rx} R^2} \tag{5.4}$$

where

P_{tx} = transmitted beam power at the wavelength of 1550 nm
η = link loss due to atmospheric turbulence and scattering
M = the all-optical modulation depth
A_s and A_r = receiving and retro-modulated areas, respectively
Ω_{tx}, Ω_{rx} = transmitted and retroreflected divergence solid angles, respectively
R = communication range between the active transceiver and the passive transceiver
    (for example, located at the ground station)

A concept architecture for a multidirectional communication is depicted in Fig. 5.8 and can be used for developing new FSO communication technologies for asymmetrical all-optical space/aerial/terrestrial networks. The figure shows the concept that can be well suited for communication with different platforms such as satellites (LEO, CubeSat), UAVs, HAPs, and balloons and therefore can be important to design FSO communication systems at terabit/s data rates to include remote locations. The transceivers at each end of multidirectional communication links transmit the laser beams from these active transceivers to the passive transceiver, which can be a ground station or any other fixed optical terminal. A glass sphere collects the active laser beams and focuses them through a thin film of CuO nanocrystals after which the retroreflected beams are sent back to each active transceiver. This way, duplex or two-way FSO communications can be achieved. Although FSO communications can be achieved using various methods of modulation and detection, an amplitude shift keying (ASK) modulation with phase-locked self-homodyne coherent detection schemes is appropriate for communication scenario. Furthermore, for phase-locked self-homodyne detection method, active transceiver laser beam can be used for both the signal beam and the phase-locked local oscillator beam [19].

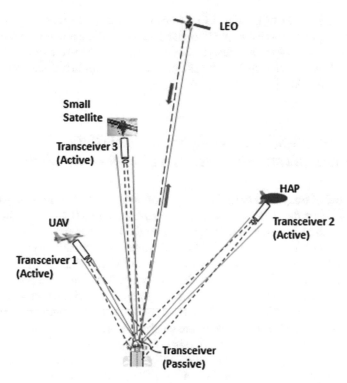

Fig. 5.8 Concept for FSO multidirectional communication system (space/aerial/HAP platforms) for achieving terabit/s using all-optical retro-modulation active/passive transceiver

5.8   New Low-Loss Plasmonic Metasurface Materials for FSO Communications Applications

A low-loss plasmonic metasurface that can collect fast-modulated light with a 3 dB bandwidth exceeding 14 GHz and a 120° acceptance angle with a record high efficiency of 30% is reported [20]. The concept technology development is based on the fact that local electromagnetic environment can be strongly modified by plasmonic structures so that the optical fields could be concentrated to deep-subwavelength regions [20]. For applications in FSO communications such as in designing emitters or detectors, it is necessary to have the core radiative properties of luminescent materials to be enhanced over macroscopic areas so that the emitters and the detectors can be integrated in the communication modules. Incoherent light sources can thus be designed using a highly efficient luminescent material and can generate ultrafast response times when modulated by a high-bandwidth laser. For FSO communication detectors with high fidelity and fast response for handling high data rate, this new type of material for macroscopic size would also be helpful to construct a large active detection area. Their research demonstrates fast modulation of the fluorescence emission with a 3 dB modulation bandwidth exceeding 14 GHz and likely approaching 80 GHz or greater. From transmitter side, this research will provide designing of a fast source compared to existing slow LEDs and thus paves the way to a new approach to on-chip design for compact, high-data-rate FSO communication system components such as ultrafast, efficient, omnidirectional detectors and incoherent sources.

5.9   Ultrahigh-Speed Free-Space Optical (FSO) Communication System Using Violet Laser Diode (VLD)

The potential of free-space optical communication beyond 25 Gbit/s using ultrahigh-speed violet laser diode (VLD) has been demonstrated for a 0.5–10 m point-to-point link [21]. In their research, the researchers used 64-quadrature amplitude modulation discrete multitone (64-QAM DMT) data stream for transmission to achieve the high data rate with received signal-to-noise ratio (SNR) of 21.34 dB and bit error rate (BER) of 3.17×10^{-3} under forward error correction. This FSO communication technology is especially applicable for short-range FSO wireless communications for indoor or turbulent/foggy outdoor environments. These short-range high-data-rate FSO communications for indoor and harsh environment are critical in establishing FSO networks integrating indoor, terrestrial, space, and underwater all-optical platforms. Visible light communication (VLC) using LED or LD and Li-Fi are extreme technologies to provide short-range FSO communications at very high data rates to meet the recent demands of high-capacity data transmission due to multimedia streaming and data exchanges among devices and networks such as 5G/6G networks. The research reported the experimental demonstration using a transistor outline can (TO-can) packaged VLD lasing at 405 nm wavelength with a threshold current of 30 mA and an ultrafast photodiode. For modulation and to evaluate the allowable transmission capacity, 64-QAM DMT was employed in their report. The demonstration is important for establishing FSO communications using compact, low-power-consuming VLD and future integration as Li-Fi in outdoor environments to establish global optical space networks.

5.10   Advanced High-Speed Photonic Technologies, Devices, and Photonic-Based Capabilities for Free-Space Optical (FSO) Communication

This section addresses some of the advanced photonic technologies in improving the speed, efficiency, and security for FSO laser communication systems in a variety of commercial, mobile, and military platforms. There has been a substantial progress in the field of FSO with the advances in high-speed optical components, subsystems, and photonic integration architectures. A variety of materials are used in photonic devices to expand the wavelength range of operation and improve performance. Some of the photonic technologies include (a) ultrafast electro-optical switches, (b) optical devices in silicon photonics, and (c) optical orthogonal frequency division multiplexing (OFDM) *ultrafast electro-optical switches for potential FSO communication application.*

Because of the exponential growth in the data volume transmitted through the optical telecommunication infrastructure, it is becoming essential for the electro-optical devices to provide optimized performance for high-spectral-efficiency transmission of information with low latency. In applications like cloud computing and Internet of Things (IoT), optical links therefore require electro-optical space switches to be able to perform ultrafast switching in a wide bandwidth. Optical amplifier-based switches demand ultrafast transition times between operational states together with reduced guard times. For linear amplification, the conventional, erbium-doped fiber amplifier (EDFA) has been widely used. Basically, SOA is a semiconductor laser without a resonant optical cavity. Because semiconductor optical amplifier (SOA) has nonlinear characteristics due to carriers' short lifetime, it is less preferable for linear amplification where EDFA has the superior performance. However, SOA's inherently nonlinear characteristics make it attractive for a number of applications such as optical switching, wavelength conversion, and multiplexing. SImultaneous amplification of 250 channels modulated at 64-QAM over an optical bandwidth of 100 nm for added ultra-wideband amplification of 115.9 Tbit/s [23]. Fast optical switching will play a very important role in providing an optical device solution for satisfying the demands of tremendous increasing data traffic including video streaming. SOA as an optical switch will have large switching capacity and low latency for optical interconnects in data centers. Experimental results discussed in [23] show the possibility of designing and building electro-optical switches with ultrafast transition time below 200 ps, low guard time below 1 ns, and moderate energy consumption up to 5 nJ/8 ns switching cycle. Some of the possible applications discussed in their report include SOA-based $M \times N$ electro-optical switches which will have direct application as key components in data centers with high switching rates and relevant to FSO communication links.

5.10.1 Optical Devices in Silicon Photonics: Passive and Active Components

Excellent research on the development of passive and active components for silicon photonic integrated circuits is reported [24]. For high-speed FSO communication systems design to meet the tremendous growing demand for data traffic. The development of these optical devices is extremely important for constructing the basic building blocks for coherent transceivers and receivers in silicon photonics. Advanced ongoing research in this area should help in designing compact, energy-efficient FSO communication architecture to address size, weight, and power

(SWaP) challenges. For the future *all-optical* free-space optical wireless communication systems, especially space (satellite)-based terminals, the development of ultracompact and integrated coherent optical transceivers is needed to accomplish broadband global Internet connectivity. Integrated optical devices will therefore eventually replace the existing discrete components for aerial, UAV, HAP, and satellite communication terminals. Photonic integrated circuit (PIC) built-in silicon-on-insulator (SOI) platform enables monolithic integration of optical and electrical devices because of the compatibility with the complementary metal oxide semiconductor (CMOS) technology and associated manufacturing process. The fundamental physics for achieving the photonic integrated circuit lies in the fact that there is high index contrast between the core (silicon: $n_{Si} \sim 3.46$) and cladding material (silica: $n_{SiO2} \sim 1.45$). Research development in this area is briefly summarized below [24]:

- Polarization splitter and rotor (PSR): This is a passive component and is important to provide polarization diversity in coherent transmitters and receivers by separating the polarization components for modulation and detection processes. PSRs which are based on "mode evolution" are based on two polarization mode conversions, first from TM_0 to TM_1 and then from TE_1 to TM_0, and offer a better tolerance to manufacturing errors, a broad bandwidth, and a lower insertion loss [24]. The report discusses the experimental demonstration of insertion loss of 1.12 dB, polarization conversion loss lower than 1.8 dB, and a cross talk below −5 dB in the wavelength region of 1520–1570 nm. Note that lasercom typically operates at a wavelength of 1550 nm so that the developed PSR would be very appropriate for FSO communication subsystem component. Another type of PSR is based on polarization cross-coupling using two waveguides in a simple directional coupler where the polarization rotation is achieved by phase-matching conditions between the TE_0 and TM_0 modes on both waveguides. This small device has a typical footprint of 1.9 μm × 3.7 μm, low polarization conversion loss of peak around 0.1 dB low, and insertion loss less than 0.1 dB over 120 nm at a broad bandwidth [24]. The research also discusses 90° hybrid, another component of coherent receiver design where the component provides a linear combination of two input fields at its four outputs to detect the real and imaginary parts of the signal.
- Modulator: The major challenges for developing integrated silicon photonic-based modulators are high efficiency and high electro-optical bandwidth. Conventional Mach-Zehnder modulator (MZM) used for optical communication systems is based on reverse polarization PN junctions with the traveling wave electrodes, has a large device length of typically about 3 mm, and is designed to have impedance match and low loss in order to provide a large electro-optical bandwidth. The plasma dispersion effect is the main physical phenomenon responsible for optical modulators using silicon. The details of designing silicon modulators of different types are already described in detail in [24]. Another key device for future FSO communication system is the tunable laser where a high output is required for generating CW wavelength light for the modulator and the

coherent receiver. At the same time, a narrow linewidth is also required to apply advanced multilevel modulation formats such as DP-M-QAM. Widely tunable hybrid laser III–V on Si is also described in detail in the same research report.

The optical devices in silicon photonics discussed above will play important roles in the future FSO communications, specially for space-based optical terminals where fast, ultracompact, energy-efficient components will be needed to meet the current and future demands for data traffic.

5.10.2   Optical Orthogonal Frequency Division Multiplexing (OFDM) Technique for FSO Communication Channel

Implementing optical OFDM for FSO communication: Some of the concepts and technologies involved in implementing optical OFDM are discussed [25]. Although the paper concentrated mostly on fiber-optics communication channel, the discussion and the simulated results to explain the optical technology can be extended to be applicable for FSO communication randomly distorted channels induced by atmospheric turbulence and multiple scattering. Two multiplexing techniques for optical orthogonal frequency division multiplexing, namely O-OFDM and Nyquist WDM, are addressed in their report. Interferometric methods for the all-optical processing are performed at the receiver side and at intermediate nodes. Based on the orthogonality condition for the O-OFDM principles, demultiplexing, routing, adding, and dropping of optical subcarrier methods are explained in their report. For communication system design involving increased number of subcarriers, Hillerkuss proposed using optical filters combined with optical couplers, Mach-Zehnder interferometers, phase shifters, and delay lines that are discussed as more compact structures. A number of optical comb generation techniques are explained. The utility of the techniques was demonstrated by simulated and experimental results of bit error rate (BER) versus optical signal-to-noise ratio (OSNR) showing the BER up to about 10^{-12}. The concept discussed can be applied to potential designing of high-speed FSO communication systems, which have to operate under real atmospheric turbulence conditions as well as for long ranges such as terrestrial, aerial, space (satellite), and underwater links.

5.10.2.1   OFDM Advanced Scheme: Magnitude and Wrap-Phase OFDM

For optical wireless communication, for example indoor communication, LED-based visible light communication (VLC) can be efficiently used for applications such as multiple input-multiple output (MIMO) links for indoor communications. Recently, researchers have described a new scheme, magnitude and wrap-phase OFDM scheme, for MIMO VLC systems. The LED transmitters can be used to provide high data transmission. The fundamental concept of the scheme can also be applied to short-range outdoor optical wireless communication with the recent Li-Fi technology and will also have potential applications to the upcoming 5G/6G information technology, as well as IoT requiring multiple optical transmitters and detectors. For conventional MIMO applications using light-emitting diode (LED) index modulation, there is restriction on the number of LEDs; however, the proposed magnitude and wrap-phase OFDM scheme for VLC system [26] decreases this restriction. The research described in the paper [26] is based on the conversion of complex signals into polar form with the magnitudes and wrap-up phases. Visible light communication LEDs operate typically within the wavelength range of 500–1000 nm. The optical orthogonal frequency division multiplexing (OFDM) techniques are chosen to overcome the severe intersymbol interference (ISI) encountered in a VLC system due to the selectivity of the VLC channels. The proposed magnitude and wrap-phase OFDM overcome this ISI problem in an optical wireless communication system and thus improve the average bit error rate (BER) of the system. The scheme is based on finding the maximum likelihood estimator (MLE) and the best linear unbiased estimator (BLUE) to decrease the restriction on the number of LEDs for improving BER performance in the optical wireless communication system. Basically, the magnitude and wrap-phase scheme is as follows:

M-ary quadrature amplitude modulation (M-QAM) format is used to modulate an information bitstream $N \log_2(M)$ where N is the number of the OFDM subcarriers and M is the modulation size. The complex OFDM resulting from the inverse FFT is converted into polar form with a magnitude and a wrap phase and represents transmitted vector. If P is the number of photodetectors at the receiver and L is the number of transmitting LEDs, the received signal vector over $P \times L$ MIMO VLC channel is then given by [26]

$$\mathbf{y} = \mathbf{Hx} + \mathbf{w} \tag{5.5}$$

where $\mathbf{H}$ is the MIMO VLC channel matrix and represents the channel gain of the VLC wireless link between each photodetector and each LED. Note that the transmitted vector has both magnitude and wrap-up phase for OFDM samples, and $\mathbf{w}$ is the vector of real-valued additive white Gaussian (AWG) noise with zero mean and a variance, σ_w^2. The above equation can thus be written as [26]

$$\mathbf{y} = h_1 x_{1,r} + h_2 x_{1,\theta} + \cdots + h_{L-1} x_{L/2,r} + h_L x_{L/2,\theta} + \mathbf{w} \tag{5.6}$$

The received signal-to-noise ratio (SNR) is given by [27]

$$\gamma = \frac{1}{\sigma_w^2} \xi \left(\frac{1}{P} \sum_{p=1}^{P} \sum_{l=1}^{L} h_{p,l} I \right)^2 \tag{5.7}$$

where ξ is the electrical-to-optical conversion factor (can be taken as a unity) and I is the mean optical intensity from the LEDs. The value of I in the above equation is chosen after ensuring that the magnitudes and wrap-up phases of the OFDM samples stay within the LEDs' dynamic ranges.

The results of BER performance of the proposed magnitude and wrap-up-phase OFDM technique are reported in the paper depicting the BER performance improvement up to 17.5 dB and reduction of the decoding complexity up to 96%, requiring only a half number of LEDs in comparison with the existing OFDM schemes.

5.11 Fast Steering Mirrors for Free-Space Optical Communication

Recently, a piezo-based two-axis fast steering mirror directing a laser beam into an 8x8 fiber matrix for beam control in FSO communication is described [28] in a Tech Blog by PI-USA. The device will be important to integrate in the high-capacity, high-data-rate FSO communication systems requiring to develop terrestrial, space, aerial, and underwater nodes for data transmission and reception of optical communication. One of the key enablers to deploy broadband global Internet connectivity is to develop *all-optical* Wi-Fi nodes globally to reach *anywhere, anytime* [29]. It is therefore absolutely necessary to implement vast space-based network of optically based communication nodes. The purpose of establishing is to link not only to ground stations but also to each other so that data can be routed from any global point to any other efficiently at very high data rate. This will thus address and provide potential solutions for the future *6G* information technology to utilize satellite/aerial/terrestrial terminals for a number of household devices, mobile devices such as in cars, and a number of modern applications including in smart cities. SpaceX discussed the first launches of its laser-equipped 142 satellites in a single payload for the first implementation of the company's inter-satellite laser links [28]. Some of the other laser-based interconnect approaches for suborbital applications include Google's balloon-borne Project Leon. In order to keep the laser beam exactly on the target over a very long distance such as satellite-to-satellite links, ground-to-satellite FSO communication links, or even deep space communication (e.g., to other planets), a high-speed fine steering system is required. For any FSO communications which have to operate under atmospheric turbulence and multiple scattering conditions, it is necessary to have a very highly accurate pointed laser beam at the receiver where the received laser beam is coupled into a single-mode fiber. The company PI-USA claims to provide angular resolution down to the nanoradian range with piezoelectric fast steering mirrors (FSMs) [28] which are very compact, fast, and highly accurate to minimize the disturbances and with mechanical bandwidth up to the kHz range. One of the company's basic designs of a fast steering mirror with differential piezo (PZT) actuation (push/pull) offers the additional advantages of angular insensitivity to temperature changes.

5.12 Advances in High-Power Laser Transmitters Relevant to FSO Communication Systems

5.12.1 High-Power Indium Phosphide Photonic Integrated Circuit Transmitters for Free-Space Optical Communications

High-power indium phosphide-based photonic integrated circuits have recently been developed to ensure high output power and high performance for applications in FSO communications. Two different types of transmitters are reported [30] based on an offset quantum well (OQW) and on a quantum well intermixing (QWI) novel platform. The OQW-based transmitter consists of a widely tunable laser, a high-speed semiconductor optical amplifier, a Mach-Zehnder modulator, and an output semiconductor optical amplifier demonstrating 14.5 dBm off-chip power and a data rate of 7 Gbit/s. The OQW-based transmitter has a footprint of 5.5 mm × 0.36 mm. The second one with QWI platform consists of a distributed Bragg reflector laser, a high-speed semiconductor amplifier, an electro-absorption modulator, and an output semiconductor optical amplifier demonstrating off-chip power of 19.5 dBm and a data rate of 20 Gbit/s. The QWI transmitter has a footprint of only 3.5 mm × 0.36 mm. For FSO communication system design, monolithic integration as compared to optical systems with discrete components definitely improves the FSO communication system reliability and improves the low-cost SWaP requirements especially for satellite/aerial/HAP/UAV applications. This fact has been demonstrated from the recent highly successful demonstrations of laser communications (lasercom) in space by various space agencies and by industries.

5.12.2 Co-Doped Fiber Amplifier for Delivering High-Power Laser Transmitter for FSO Applications

Free-space optical (FSO) communications between space-based and ground terminals require long-range transmission and reception of optical links. This is a typical communication scenario which is encountered between low-orbit satellites and ground transceivers. The performance of a long-range FSO communication system depends on the laser transmitter power, which also determines the quality of communication capacity and the achievable maximum data rates. Transmitting laser power is also important for broadband Internet connectivity, which will need integrating both low-orbit satellites and terrestrial optical networks. At the same time, it is important to develop the FSO communication modules and the devices to withstand the radiation in the space. Generally, an optical amplifier is used to achieve high power in a laser transmitter. A recent report [31] investigates radiation-hardened high-power Er^{3+}/Yb^{3+} Co-doped fiber amplifiers for free-space optical communications applications. In their report, high-power erbium/ytterbium Co-doped fiber amplifiers deliver 20 W of signal output power at 1565 nm. The method is based on pumping at 915 nm (43 W) radiation-tolerant (Er/Yb) or radiation-hardened (ErYbCe) active few mode fibers. The results show that the gain degradation levels of high-power Co-doped fiber amplifier based on radiation-hardened fibers are below 6%.

5.12.3   All-Fiber Master Oscillator Power Amplifier (MOPA) for High-Power Laser Transmitter

Many of the emerging free-space optical communications applications require high-power optical amplifiers, which are designed to satisfy different wavelength band, modulation, polarization, and atmospheric channel. These optical amplifiers serve as a power booster at the transmitter end whose characteristics are determined by the free-space communication link range and atmospheric channel, required data rate, optimal transmission format, and other integrable FSO devices. Because long-range FSO communication for space (satellite)-based system typically requires one watt to tens of watts power output, cheap sources of optical pump power and efficient gain media are in demand. The gain of the optical amplifier must be typically 40–55 dB to boost a modulated source to achieve these levels of power. For airborne and space vehicles, there are significant requirements for achieving efficient FSO communication performance. Finally, reliability plays an important consideration for designing high-power optical amplifier subject to extreme environments and suitable for FSO communication for long-range space links.

High-power fiber sources, as an attractive alternative to solid-state lasers, can be very useful for FSO communication laser sources operating at 1555 and 1563 nm wavelengths. A 40 W all-fiber Er/Yb master oscillator power amplifier (MOPA) has been recently demonstrated [32, 33]. The three-stages of MOPA system are as follows: the first stage for low-power amplification utilizing active single-mode single-clad (SM-SC) fibers, the second stage for medium-power amplification still based on SM-SC, and the third amplifier stage for high-power amplification for output power of dozens or more watts using a large-mode-area double-clad (LMA-DC) fiber. The low-power signal laser is modulated where power intensities are much lower. The results demonstrated an all-fiber high-power CW-operating three-stage MOPA system preamplifier (EDFA based on erbium-doped active single-mode fibers), medium-power amplifier (EYDFA-based Er/Yb Co-doped active SM-DC fiber), and high-power amplifier (EYDFA based on Er/Yb Co-doped active LMA-type DC fiber) using self-fabricated high-power passive components. The achieved power reported was 44 W at the wavelengths of 1555 and 1563 nm in the presence of pumping at the power of 144 W and the wavelength of 915 nm. The overall system can be compact and is integrable with other FSO communication devices and modules and can be useful for potential space (satellite)-based and terrestrial terminals to establish high-capacity uplink communication at very high speed.

5.13   Other Related Advances in Free-Space Optical Communication Technologies Applicable to FSO Communication for Space/Aerial/Terrestrial/ Underwater (SATU) Links

This book provides *all-optical* concept technology for establishing global Internet connectivity to have access for all. Special emphasis is given to discuss space optical links using laser satellite constellation for various orbits such as GEO, MEO, and MEO. Integrated space/aerial/terrestrial/underwater (SATU) links are essential in order to transfer data information from a sender to the destination locations worldwide. In the all-optical concept technology, the terminals in space, terrestrial, or underwater platforms need to be equipped with laser/optical transceivers at each end. Some of the advanced technology schemes are discussed earlier. Some of the current advanced technology researches relevant to all-optical Internet communication and connectivity are discussed in the author's book [34]. Some of the technologies useful for developing satellite, terrestrial, and underwater optical communication links are listed below:

1. *Plasmonic device for 100 Gbit/s optical links*
   A microscale modulator using plasmonically active gold components is reported which demonstrates fast operation for ultra-broadband signals. The modulators operating greater than 100 Gbit/s limit of photonic devices for a single carrier were presented at the OFC Conference in Los Angeles. The research work clearly paves a new way to meet today's high demands for the Internet traffic over optical networks for optical communication.
2. *High-capacity, ultra-broadband on-chip twisted light emitter for optical communication systems*
   An ultra-broadband multiplexed orbital angular momentum (OAM) emitter for OAM communication with a data rate of 1.2 Tbit/s assisted by 30-channel optical frequency combs (OFCs) is reported [35]. This has potential application in very-high-speed, high-capacity FSO communication system, which can also provide an additional degree of freedom for designing wavelength/frequency division multiplexing. The wavelength of the emitter covers 1450–1650 nm wavelength, which falls at the typical lasercom wavelength of 1550 nm.
3. *Optical phase-sensitive amplifiers (PSAs)*
   The research project at the Chalmers University of Technology, Gothenburg, Sweden, developed optical amplifiers, which can achieve a quantum-limited noise figure (NF) of 0 dB for a noiseless amplifier [36]. This advanced technology will be extremely useful in space optical links for satellites and aerial terminals.

4. *Compact optical transceiver by hybrid multichip integration*

A recent US patent [37] discusses a 3D hybrid multichip stacking without wire bonds and method. The compact optical transceiver will find a number of applications in the practical design of developing optical communication technology for the space-based system for decreased footprint.

5. *All-optical routing and switching for 3D photonic circuitry*

For space/aerial/terrestrial optical links in FSO communication systems, routing and switching are absolutely necessary to complete a two-way communication. The research reported an experimental demonstration of all-optical routing and switching for a 3D photonic circuitry, which will be extremely important in developing future communication system to support three-dimensional network topologies for photonic circuitry with the potential possibility of expanding the switch port to exceed 1000 × 1000 port counts for data center applications [38, 39].

Some of the research advances will be essential in successful demonstrations of optical wireless communications between LEO, small, CubeSat, high-altitude platform (HAP), balloon, and ground terminals as well as between satellites (ISL) and between HAP/balloon/small satellites.

5.14 Discussions

This chapter discusses current advanced research relevant to "all-optical" communications and connectivity and future directions to make the concept of laser satellite constellation for achieving very-high-data-rate, high-capacity communication systems successful. This chapter presents various technology developments, research challenges, and advances in FSO communication system as well as discusses research challenges to be addressed and solved to overcome the limitations of existing systems and emerging new technologies. The recent applications related to FSO wireless communication systems also include space (aerial)-based and underwater platforms, 5G/6G networks, smart cars, robots, optical links for aiding health-related services, and artificial intelligence (AI)-based machine-to-machine (M2M) devices for adaptive optics (AO).

References

1. SyFy Wire, Feb. 18, 2020.
2. Kamugisha KAZAURA, Kazunori OMAE, Toshiji SUZUKI, Mitsuji MATSUMOTO, Edward MUTAFUNGWA, Tadaaki MURAKAMI, Nonmember, Koichi TAKAHASHI, Hideki MATSUMOTO, Kazuhiko WAKAMORI, and Yoshinori ARIMOTO, "Performance Evaluation of Next Generation Free-Space Optical Communication System," IEICE TRANS. ELECTRON., VOL., E90-C, NO. 2 FEBRUARY 2007.

3. Suzana Miladić-Tešić, Goran Z. Markovic, Dragan Perakovic and Ivan Cvitić, "A review of optical networking technologies supporting 5G communication infrastructure," DOI: https://doi.org/10.1007/s11276-021-02582-6

4. Elias Yaacoub and Mohamed-Slim Alouini, "A Key 6G Challenge and Opportunity Connecting the Remaining 4 Billions: A Survey on Rural Connectivity," arXiv:1906.11541v1, 27 Jun 2019.

5. Pirinen, "A brief overview of 5G research activities," in Proc. of International Conference on 5G for Ubiquitous Connectivity (5GU), Nov. 2014, pp. 17-22.

6. Mostafa Zaman Chowdhury, Md. Tanvir Hossan, Amirul Islam, and Yeong Min Jang, "A Comparative Survey of Optical Wireless Technologies: Architectures and Applications," https://arxiv.org/ftp/arxiv/papers/1810/1810.02594.pdf

7. Mostafa Zaman Chowdhury, Md. Shahjalal, Moh. Khalid Hasan and Yeong Min Jang, "The Role of Optical Wireless Communication Technologies in 5G/6G and IoT Solutions: Prospects, Directions, and Challenges," Appl. Sci. 2019, 9, 4367; doi:https://doi.org/10.3390/app9204367, 16 October 2019.

8. MOSTAFA ZAMAN CHOWDHURY, MD. SHAHJALAL, SHAKIL AHMED AND YEONG MIN JANG, "6G Wireless Communication Systems: Applications, Requirements, Technologies, Challenges, and Research Directions," IEEE Open Journal of the Communications Society, Digital Object Identifier https://doi.org/10.1109/OJCOMS.2020.3010270, 4 August 2020.

9. Artem M. Vorontsov, Mikhail A. Vorontsov, Grigorii A. Filimonov and Ernst Polnau, "Atmospheric Turbulence Study with Deep Machine Learning of Intensity Scintillation Patterns," Appl. Sci. 2020, 10, 8136; doi:10.3390/app10228136, 17 November 2020.

10. Ravikiran Kakarla, Jochen Schröder and Peter A. Andrekson, "One photon-per-bit receiver using near-noiseless phase-sensitive amplification," Light: Science & Applications (2020) 9:153, https://doi.org/10.1038/s41377-020-00389-2

11. Sensitive Optical Receivers for Space, Tech Briefs, February 1, 2021.

12. Peter Andrekson, Chalmers University of Technology, *Private Communication* (via email), October 2020.

13. ALEXANDER D. GRIFFITHS, JOHANNES HERRNSDORF, ROBERT K. HENDERSON, MICHAEL J. STRAIN AND MARTIN D. DAWSON, "High-sensitivity inter-satellite optical communications using chip-scale LED and single-photon detector hardware," Optics Express, Vol. 29, No. 7/29 March 2021/Optics Express 10749.

14. "World record in free-space optical communications", https://www.dlr.de/content/en/articles/news/2016/20161103_world-record-in-free-space-optical-communications_19914.html

15. PhotonicsViews:Optics-Photonics-Laser Technology, "World Record in Free-Space Optical Communications," 7th November 2016.

16. Juraj Poliak, Ramon Mata Calvo, and Fabian Rein, "Demonstration of 1.72 Tbit/s Optical Data Transmission Under Worst-Case Turbulence Conditions for Ground-to-Geostationary Satellite Communications," IEEE COMMUNICATIONS LETTERS, VOL. 22, NO. 9, SEPTEMBER 2018.

17. E. Ciaramella, Y. Arimoto, G. Contestabile, M. Presi, A. D'Errico, V. Guarino, and M. Matsumoto, "1.28 Terabit/s (32x40 Gbit/s) WDM Transmission System for Free Space Optical Communications," IEEE Journal on Selected Areas in Communications, Vol. 27, No. 9, December 2009.

18. Arun K. Majumdar, *Advanced Free Space Optics (FSO): A Systems Approach, Springer*, New York 2015.

19. Brandon Born, Ilija R. Hristovski, Simon Geoffroy-Gagnon, and Jonathan F. Holzman, "All-optical retro-modulation for free-space optical communication," Optics Express, Vol. 26, No. 4 | 19 Feb 2018 | OPTICS EXPRESS 5031.

20. Andrew J. Traverso, Jiani Huang, Thibault Peyronel, Guoce Yang, Tobias G. Tiecke, and Maiken H. Mikkelsen, "Low-loss, centimeter-scale plasmonic metasurface for ultrafast opto-electronics," OPTICA, Vol. 8, No. 2 / February 2021.

21. Wei-Chun Wang, Huai-Yung Wang & Gong-Ru Lin, Ultrahigh-speed violet laser diode based free-space optical communication beyond 25 Gbit/s, Scientific Reports, September 2018.

22. Sanjaya Lohani, Erin M. Knutson & Ryan T. Glasser, Generative machine learning for robust free-space communication, Communication Physics, 3:177 (2020). https://doi.org/10.1038/s42005-020-00444-9
23. Tiago Sutili, Rafael Carvalho Figueiredo, Bruno Taglietti, Cristiano M. Gallep and Evandro Conforti, Ultrafast Electro-Optical Switches Based on Semiconductor Optical Amplifiers, Springer Nature Switzerland AG 2019, A. Paradisi et al. (eds.), *Optical Communications, Telecommunications and Information Technology,* https://doi.org/10.1007/978-3-319-97187-2_2
24. Yesica R. R. Bustamante, Uiara Moura, Henrique F. Santana and Giovanni B. de Farias, Optical Devices in Silicon Photonics, Springer Nature Switzerland AG 2019, A. Paradisi et al. (eds.), *Optical Communications, Telecommunications and Information Technology,* https://doi.org/10.1007/978-3-319-97187-2_11
25. Mônica L. Rocha, Rafael J. L. Ferreira, Diego M. Dourado, Matheus M. Rodrigues, Stenio M. Ranzini, Sandro M. Rossi, Fabio D. Simões and Daniel M. Pataca, Challenges Toward a Cost-Effective Implementation of Optical OFDM, Springer Nature Switzerland AG 2019, A. Paradisi et al. (eds.), *Optical Communications, Telecommunications and Information Technology,* https://doi.org/10.1007/978-3-319-97187-2_8
26. Mohamed Al-Nahhal, Ertugrul Basar, and Murat Uysal, Magnitude and Wrap-Phase OFDM for MIMO Visible Light Communication Systems, DOI 10.1109/LCOMM.2021.3070272, IEEE Communications Letters, 1089-7798 (c) 2021 IEEE.
27. A. Yesilkaya, E. Basar, F. Miramirkhani, E. Panayirci, M. Uysal, and H. Haas, "Optical MIMO-OFDM with generalized LED index modulation," IEEE Trans. Commun., vol. 65, no. 8, pp. 3429–3441, Aug. 2017.
28. PI-USA, Tech Blog, Fast Steering Mirrors for Deep Space / Free Space Optical Communication, MAY 2021 https://www.pi-usa.us/en/tech-blog/
29. Arun K. Majumdar, Optical Wireless Communications for Broadband Global Internet Connectivity: Fundamentals and Potential Applications, Elsevier, Amsterdam, Netherlands, 2019.
30. Hongwei Zhao, Sergio Pinna, Fengqiao Sang, Bowen Song, Simone Tommaso ˇSuran Brunelli, Larry A. Coldren, and Jonathan Klamkin, "High-Power Indium Phosphide Photonic Integrated Circuit," IEEE Journal of Selected Topics in Quantum Electronics, Vol 25, No. 6, November/December 2019.
31. Ayoub Ladaci, Sylvain Girard, Luciano Mescia, Arnaud Laurent, Carine Ranger, David Kermen, Thierry Robin, Benoit Cadier, Mathieu Boutillier, Baidy Sane, Emmanuel Marin, Adriana Morana, Youcef Ouerdane, and Aziz Boukenter, "Radiation hardened high-power Er^{3+}/Yb^{3+}-codoped fiber amplifiers for free-space optical communications," Optics Letters, Vol. 43, Issue 13, pp. 3049-3052 (2018).
32. L. Stampoulidis, J. Edmunds, M. Kechagias, G. Stevens, J. Farzana, M. Welch and E. Kehayas, "Radiation-resistant optical fiber amplifiers for satellite communications", Proc. SPIE Vol. 10096, 100960H, 2017.
33. Pawel Kaczmarek 1, Dorota Stachowiak 2,* ID and Krzysztof M. Abramski, "40 W All-Fiber Er/Yb MOPA System Using Self-Fabricated High-Power Passive Fiber Components," Appl. Sci. 2018, 8, 869.
34. Arun K. Majumdar, "*Optical Wireless Communications for Broadband Global Internet Connectivity: Fundamentals and Potential Applications,*" Elsevier, Amsterdam, Netherlands 2019.
35. Z. Xie, T. Lei, F. Li,H.Qiu, Z.Zhang, H. Wang, C. Min, L. Du, Z. Li and X. Yuan, "Ultra-broadband on-chip twisted light emitter for optical communications," Light: Science & Applications (2018) 7, 18001; doi:10.1038/lsa.2018.1; published online 20 April 2018.
36. Chalmers University project, "Noiseless phase-sensitive optical amplifiers and their applications," Published: Fri 29 Jan 2016: Fri 15 Sep 2017.

37. Patent number: 9921379, Date of Patent: Mar 20, 2018, Patent Publication Number: 20170261708, Assignee: INPHI CORPORATION (Santa Clara, CA) Inventors: Liang Ding (Singapore), Radhakrishnan L. Nagarajan (Santa Clara, CA), Roberto Coccioli (Westlake Village, CA).
38. R. Keil, M. Heinrich, F. Dreisow, T. Pertsch, A. Tunnermann, S. Nolte, D. N. Christodoulides and A. Szameit, "All-optical routing and switching for three-dimensional photonic circuitry," SCIENTIFIC REPORTS, 1: 94 | DOI:10.1038/srep00094, 15 September 2011.
39. K. Sato, H. Hasegawa, T. Niwa, and T. Watanabe, "A Large-Scale Wavelength Routing Optical Switch for Data Center Networks," IEEE Communications Magazine, September 2013.
40. Sanjay Lohani, Erin Knutson and Ryan Glasser, Generative machine learning a for robust free-space communication, Communication Physics 2020.

Chapter 6
Constellation of Satellites: Integrated Space/Aerial/Terrestrial/Underwater (SATU) All-Optical Networks for Global Internet Connectivity

6.1 Introduction

This chapter introduces the major purpose of writing this book, which is to develop design concept technology leading to high-bandwidth, high-data-rate global Internet connectivity and seamless communications using all-optical technology based on integrating laser/optical communication constellation satellites, UAVs, balloons, and aircraft. The chapter starts with a comparison between single, conventional, and constellation of satellites and addresses the issues relevant to free-space optical communications such as various locations of constellation satellites at different orbits (e.g., minimizing latency in satellite networks), *handover* and *switchover* problems and solutions, different atmospheric propagation effects for each satellite LOS propagation link, and maintenance of uninterrupted/seamless communications. The analysis will discuss the optimum number of satellites in a given orbit for best communication performance, different communication links for the constellation satellites, and various types of satellites involved in communication constellation: conventional communication satellites, small satellites, microsatellites, CubeSats, and picosatellites: specific orbits: optimum number of satellites for a given orbital heights, data rates, and minimizing of latency in the network.

Why a single satellite cannot establish global wireless Internet connectivity anytime, anywhere?

– A line-of-sight communication link from a terrestrial location equipped with transceiver device toward a single satellite is required for any position of the satellite in the orbit, and therefore it is not possible to maintain the LOS link. The receiver antenna (RF or optical) FOV has to accommodate the transmitting beam for needed received signal for communication.

Why a constellation of satellites is required for connecting to any location on the Earth from a satellite in the constellation?

© Springer Nature Switzerland AG 2022
A. K. Majumdar, *Laser Communication with Constellation Satellites, UAVs, HAPs and Balloons*, https://doi.org/10.1007/978-3-031-03972-0_6

– Even for a single Earth location, more satellites are required to avoid cloud and to maintain two-way communication links. Coordinates and coverage geometry for any location of a satellite and the ground are necessary. A number of ground stations located in various places on the Earth are required to send and receive *two-way* communication from any location to another location, worldwide. A constellation of satellites can provide continuous, global coverage as the satellite moves.

This chapter also addresses why laser FSO link is needed for high-speed, high-capacity communication to satisfy today's demand for enormous data transfer. This section addresses the issues of constellation and laser FSO link for establishing global communication.

6.2   Need to Develop Technology Concept for Establishing High-Speed Global Broadband Internet Connectivity in Remote Places

The motivation and purpose for writing this book specially to address the technology developments for providing wireless communications and Internet connectivity in remote locations started when the author was an invited panelist at the Optical Society of America (OSA) Advanced Photonics Congress Symposium, San Francisco (July 30, 2019), with the title "A Light in Digital Darkness: Optical Wireless Communications to Connect the Unconnected." Although the "Global Internet Connectivity" or "Connecting the Unconnected" theme is becoming an important topic from different perspectives, it is typically ignored in major OSA, SPIE, and IEEE conferences where most of the focus is on 5G and beyond 5G networks for highly dense urban environments. Global access to Internet connectivity has the potential of providing access for all. Free-space optics (FSO) is a practical solution for creating a three-dimensional global communication grid. FSO technology is a powerful tool to address connectivity bottlenecks and can allow worldwide access to the Internet independent of terrestrial limitations by establishing wireless links through air and satellite communications providing access to fixed and mobile services. The complete solution should handle indoor, outdoor, terrestrial, and space links for successful secure global Internet connectivity. There have been recent advancements in putting many satellites in constellations, and the advanced free-space optics (FSO) communication technology is just about the right technology to provide the best technical solution. This book addresses the needs to provide innovative technology concepts integrating both space and terrestrial links using all-optical based FSO technology to provide broadband Global Internet connectivity and communications *anywhere, anytime*.

6.3   Applications of Free-Space Optical (FSO) Communications: Space Optical Information Networks

Why to connect small satellites to Earth and each other with laser link?

Laser links would provide the backbone (similar to very-high-speed fiber for terrestrial and long-haul submarine) for establishing a global wireless network. Low Earth orbit satellite constellations and digital connectivity will provide global communications and connectivity worldwide. This section discusses the importance of FSO communication technologies to develop space optical communication and information networks. Free-space optical (FSO) communication and space optical information network can offer the demand of tremendous increased data transfer as well as to establish in distant remote locations. To develop and design global coverage concept will need to create the primary hubs at the equinox data centers where the bulk of the world's networks interconnect. These primary hubs can distribute data from space to terrestrial networks at various locations. The basic concept will be to use a number of ground nodes comprising each hub equipped with optical/laser heads to link with a number of different optical/laser satellites belonging to a constellation [2]. *This way, there will always be a line-of-sight (LOS) optical link connectivity available with any of the satellites in that constellation at that location and time.* This concept design approach can offer more advantages over optical fiber, RF, or microwave communication technologies such as smaller (SWaP) terminals (very important for satellite-based laser/optical transceivers), portability, flexibility, and lower costs. Finally, instant high-speed data transfer between laser/optical satellites (or HAPs, balloons, UAVs) and any remote locations on the Earth can be established *anytime, anywhere*. Figure 6.1 shows the example of how laser satellite in a constellation will be interconnected with ground nodes and with each other. The space terminals can also include HAPs, balloons, and UAVs as mentioned before. Also, submarine high-speed optical cable landing stations can be included to other terrestrial node locations so that it will be possible to transfer data across the oceans and then to distribute almost anywhere on the ground. Future developments of data rates of 100 Gbit/s between ground nodes and satellites and 200 Gbit/s from satellite to satellite are predicted [3]. The data rates are about 100 times or faster than radio links used in satellite communication today.

Because of the vast advancement in optical technology, fundamental changes are occurring in the way global communication network infrastructure is designed. The future global network backbone will consist of intercontinental submarine fiber cables capable of handling unprecedented amounts of data bandwidth, and the constellation of optical satellites and other space terminals such as HAP, balloons, and UAVs beaming data to the Earth using lasers. This backbone is therefore essential to provide global connectivity to anywhere on the Earth including any remote places to enable a lot of data from centralized computing hubs. Artificial intelligence (AI) can be used to route data packets around cloud so that communication is not interrupted because of clouds in the LOS path, and also SD WANs can be deployed to manage its global space network. In short, the space optical information networks

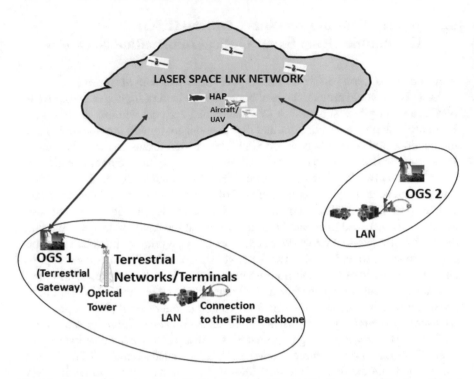

Fig. 6.1 Basic architecture of space/Earth integrated optical network using satellites/space/aerial and terrestrial terminals. The link network can be achieved with a single and a number of optical satellites in a constellation and/or HAP/balloons, UAVs, and terrestrial terminals

and ground communication networks can be integrated to construct space/Earth high-speed information network.

6.3.1   Basic Architecture of Space Optical Communication Networks

A very basic, high-level space optical network using satellites/space/aerial and terrestrial terminals is depicted in Fig. 6.1. The link network in the figure can be achieved with a single or a number of optical/laser satellites in a constellation and/or HAP/balloons, UAVs, and terrestrial terminals. The basic architecture includes (1) a single satellite and/or satellites belonging to a constellation at a given orbital height forming a high-speed space/Earth integrated communication network; (2) ground laser/optical relay stations acting as forwarding *nodes* between the ground communication systems and space-based satellite optical communication network, an example of which is ground mobile terminals to establish space/ground integrated communication networks; and (3) point-to-point laser communication

ground terminals for short ranges requiring very-high-speed, high-capacity communication. The integrated space/Earth architecture to achieve high-speed dynamic
performances in real time is based on optical/laser space communication technology applied to various platforms located in both terrestrial and in different altitudes
(determined by the orbits). FSO laser communications depend mainly on both forward and backward information links. In case of forward link, the information
received at the optical ground stations from different ground laser communication
relay stations is transmitted to a satellite as *uplink* in the satellite/ground architecture. Similarly, the received optical signals at the satellite can be retransmitted as
downlink via either inter-satellite or a satellite-to-HAP/balloon/UAV communication to another location served by another ground station so that high-speed FSO
bidirectional information transmission can be accomplished combining both uplink
and downlink transmissions.

6.3.2   Constellation of Satellites for Laser Space Communications Network

Laser/optical links from satellites or space terminals to the ground have been receiving significant interest in recent years. Increasing the carrier frequency from the
radio spectrum to the optical wavelength using advanced lasers around 1.55 μm has
the advantage of providing very high data rates and also of offering greater security
against eavesdropping by the small print of the signal on the ground. Figure 6.2
shows an artist's concept of GEO-LEO-ground terminals in a laser space network
for establishing FSO communications between two ground locations on the Earth.
Free-space optical wireless communication technologies have been constantly
developed for providing compact, portable high-speed communication devices.
Some of these recent developments are the following:

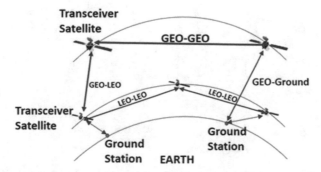

Fig. 6.2 Artist's concept of GEO-LEO-ground terminals in a laser space network for establishing
FSO communications between two ground locations on the Earth

- High-power diode lasers in the 1550 nm lasercom range—high-power laser chips, Fabry-Perot laser chips, SFB lasers in 1310–2000 nm, 1550 nm laser diode, high-power 20 W fiber-coupled module
- Fast detectors—nanosecond free-space optical detectors, single photon detector
- Optical chips to perform complex onboard signal processing
- Advanced acquisition pointing and tracking (APT) modules (MEMS based and also with no moving parts)
- Efficient modulation techniques for high communication capacity
- Atmospheric turbulence mitigation methods including real-time adaptive optics (AO): generative machine learning such as generative neural network (GNN) to control long-range FSOC links
- Silicon photonics (to enable faster data transfer over long distances), integrated photonics circuit (PIC) technologies to achieve extremely small size and weight transceivers
- Potential integration of FSOC for 5G/6G and IoT applications and artificial intelligence (AI) and machine language (ML) to control FSOC

A potential integrated *all-optical* satellite/terrestrial network is shown in Fig. 6.3 that includes both LEO satellites and relay optical links. The basic space communication architecture for a potential integrated *all-optical* satellite/terrestrial network includes the following:

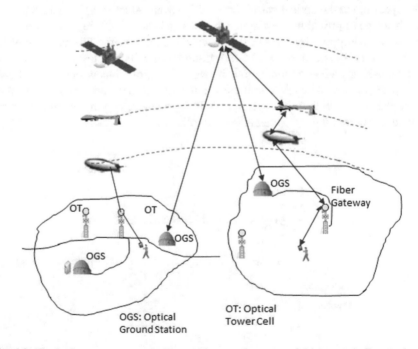

Fig. 6.3 The basic space communication architecture for a potential integrated *all-optical* satellite/terrestrial network. (*Reprinted with permission from the publisher Elsevier, Amsterdam, Netherlands 2021* [2])

6.3.2.1   Low Earth Orbit (LEO) Satellite Equipped with Laser/Optical Transceiver Network

In order to achieve global coverage using laser-equipped satellites, there are two basic (and practical) parameters needed: (1) minimum number of satellites required in a constellation which also depends on the orbital altitude and the locations of the satellites and (2) available locations of the optical ground stations (OGS) on the Earth. The number of satellites required must be determined from FSOC parameters such as the laser-transmitting beam divergent angle and the receiver field of view (FOV). In any specific real-time dynamic communication network, an accurate design of FSO communications system is necessary. Typically, the footprint of the signal (spot size on the ground) is only a few meters in diameter at distances from the low Earth orbit (LEO). Generally, about 60–80 satellites will be needed to achieve global coverage [4]. The LEO laser satellites in a given constellation can first establish the LEO-LEO inter-satellite communication link with an adjacent LEO satellite and then a LEO-OGS communication link from a OGS at a given location on the Earth. The LEO satellite laser communication can thus be connected to the relay satellite communication network. *Finally, in principle, it is entirely possible to establish a bidirectional laser communication link between two users in the following way: one user in one location served by a particular OGS can connect another user located in another location served by another OGS on the Earth.*

6.3.2.2   Relay Satellite Communication Network Using Constellation

A relay laser/optical communication network can be achieved in a number of ways as follows. Some of the possible scenarios are:

- Inter-satellite and interorbital (GEO-GEO, LEO-LEO, GEO-LEO, and LEO-GEO) network consisting of satellites in different orbits
- LEO1 to OGS1 (LEO # 1 is line-of-sight (LOS) link with OGS1 at a particular location on the Earth), LEO1 to LEO2, and then LEO2 to OGS1 as long as the laser-transmitting divergent angles from both LEO1 and LEO2 can accommodate the FOV of the OGS1
- LEO-GEO link and GEO to OGS link
- GEO-GEO link

Although Fig. 6.3 shows space network links only with GEO and LEO satellite terminals, the complete and comprehensive space/aerial network will also include small, micro, and cube satellites; HAPs; balloons; UAVs; and aircraft. In addition, for global connectivity, underwater laser communication links must be integrated for a complete space/aerial/underwater communication network. All the links between GEO and GEO as well as between LEO and LEO satellites must be designed and built according to the most efficient protocol, and laser signals transmitted between the relay satellites will be accomplished by chain transmission with proper *handover* schemes. Space optical networks with both GEO and LEO inter-satellite communication network can thus be established. Users can thus access both the GEO satellite communication network and the LEO networks.

Various FSO networks involving laser/optical satellites, terrestrial, and LAN networks can be integrated to establish *all-optical* broadband global connectivity in any location [2]. An optical ground station (OGS) eventually establishes connectivity between the laser satellites belonging to a constellation and the OGSs as shown in Fig. 6.4. Optical space network integrating land, air, and space environments, which include a constellation of satellites (and other platforms such as HAPs, UAVs, balloons), is already discussed earlier.

A general concept architecture showing high-altitude platforms (HAPs) integrated with satellite and terrestrial systems is depicted in Fig. 6.5, which also includes inter-platform links (IPLs).

6.3.2.3   Multilayer Space Communication Architecture

Because of the recent technological advance of the aerial and miniaturized satellite platforms, FSO wireless communications can play an important role in intermediate layers of communication systems between terrestrial and traditional satellite segments. Depending on the operation altitude, these new platforms in the intermediate layers can be classified accordingly for various applications. Some of the categories are:

(a) Very low Earth orbit (VLEO) satellites: altitude range ~100–450 km
(b) High-altitude platforms (HAPs) and balloons: altitude range ~15–25 km (HAPs) and ~18–25 km (high-altitude balloons, e.g., Project Loon)
(c) Low-altitude platforms (LAP): altitude range ~0–4 km

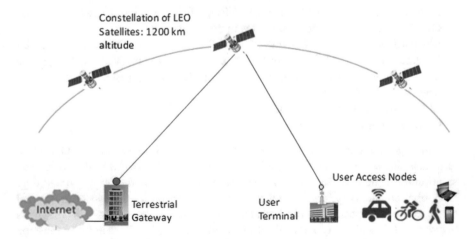

Fig. 6.4 A simplified architecture for establishing global communication networks using a constellation of satellites. (*Reprinted with permission from the publisher Elsevier, Amsterdam, Netherlands 2021* [2])

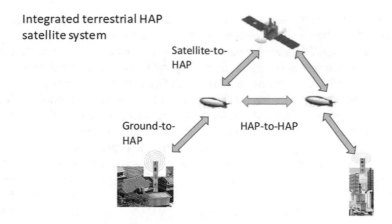

Fig. 6.5 High-altitude platforms (HAPs) integrated with satellite and terrestrial systems, which also includes inter-platform links (IPLs). (*Reprinted with permission from the publisher Elsevier, Amsterdam, Netherlands 2021* [2])

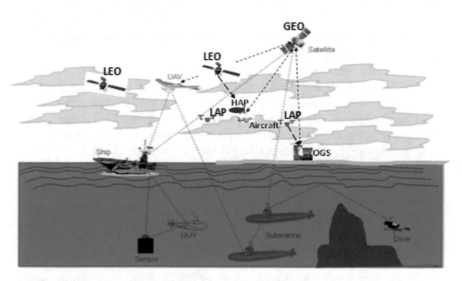

Fig. 6.6 Schematic diagram and an architecture for multilayer space communication network that includes satellite-to-submarine blue-green laser communication link

Figure 6.6 shows a schematic diagram and an architecture for multilayer space communication network that includes GEO, LEO, HAP, LAP, and also satellite-to-submarine blue-green laser communication links.

In the multilayer architecture approach, the VLEO and satellite communication-assisted aerial networks can also be integrated to overcome the most challenging scenarios. Platforms at various altitudes offer both benefits and some challenges.

Very low Earth orbit platforms operate at altitudes closer than the typical LEO satellites and can be smaller, cheaper, and simpler. However, VLEO platforms face more aerodynamic forces because of more atmospheric density at the lower altitudes and therefore may require more frequent spacecraft replacements.

High-altitude platforms can act similar to satellite networks but have the potential to establish communication services to local regions on the ground. HAPs can serve as an intermediate communication element connecting the two-communication links and can be used as communication *backhaul*: satellite-to-HAP and HAP-to-ground links. Atmospheric turbulence and scattering effects for satellite-to-ground long communication path are much more severe than for a shorter HAP-to-ground path. FSOC system design requiring laser/optical transceiver components on the platforms depends ultimately on the link budget, which is much improved for the shorter range. HAPs offer extreme flexibility in the network; for example, the users within a specific regional coverage area can be directly connected as well as with links with satellite for inter-coverage communications. Some of the challenges for HAPs include reduced available daylight hours, dragging of the HAPs due to high wind speed, and limited lifetime of batteries. One of the main advantages of the HAPs is the lower propagation delay of ~50–85 μs compared to GEO (~120 ms), MEO (~15–85 ms), and LEO satellites (~1.5–3 ms) [5].

Low-altitude platforms include unmanned aerial vehicles (UAVs) and tethered balloons. Recent advances in UAV technologies make them a good candidate for an element in a communication network specially at a cellular level for fast and flexible deployment, for establishing direct line-of-sight (LOS) communication link. Other great advantages of UAVs are their capabilities as autonomous and controlled mobility communication platforms. To make UAVs a practical and viable part of the network element, safety features of the drones to avoid collisions with other neighboring objects, security, and public acceptances need to be considered.

6.3.2.4   OGSs Providing Ground User Terminal Links

This section discusses briefly various types of FSO systems for different network architectures, which integrate satellite communication, free-space optics (terrestrial), and wireless LAN combined. Free-space optical wireless communication has the potential solution to satisfy the tremendous demand for high-bandwidth and secure communication between various devices such as smartphones, PCs, and laptops within a communication network served by inter-building/terrestrial, aerial, space (HAPs, UAVs, and balloons), and satellite links. Figure 6.7 depicts a simplified diagram showing various FSO links involving laser/optical satellite, terrestrial, and LAN integrated with a fiber-optic backbone and the Internet server. The FSO-based LAN will provide the connectivity to fixed and mobile users within a network cell, whereas the terrestrial FSO will establish broadband data link between two

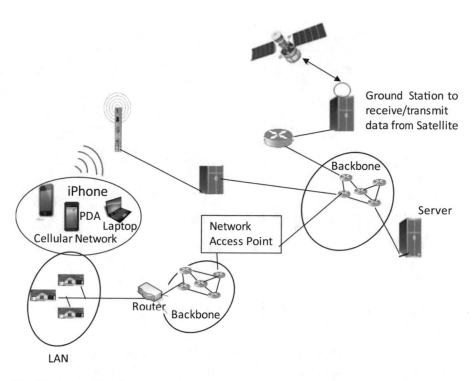

Fig. 6.7 A simplified diagram showing various FSO links involving laser/optical satellite, terrestrial, and LAN integrated with a fiber-optic backbone and the Internet server. (*Reprinted with permission from the publisher Elsevier, Amsterdam, Netherlands 2021* [2])

fixed locations providing the broadband wireless solution for closing the "*last-mile*" connectivity gap throughout metropolitan networks.

For achieving successful *all-optical* global network, some of the important issues to consider are the reliability and the availability of FSO links and FSO networks [6]. Reliability of each individual components such as all-electrical and -photonic components built into the terminal as well as the connection and interface to the network need to be taken into account. Figure 6.8 shows an artist's concept of establishing *all-optical* global communications and Internet connectivity between two countries, e.g., the USA and Europe.

6.4 Concept Designs of Laser Satellite Constellations

This section discusses some system aspects of hierarchy and high-level laser communication system architecture. Note that satellites such as GEO and LEO and other space platforms such as HAPs, UAVs, and air platforms are still not equipped with laser/optical transceivers. In addition, both inter-satellite links (ISLs) and

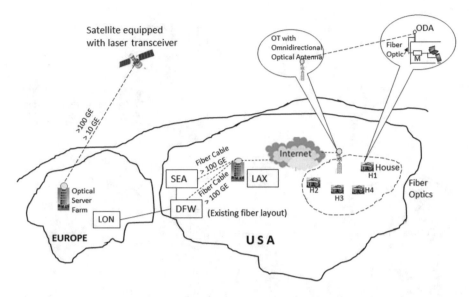

Fig. 6.8 An artist's concept of establishing *all-optical* global communications and Internet connectivity between two countries

inter-platform links (IPLs) are also not equipped with all-optical subsystems and components such as adaptive optics (AO) and advanced acquisition, tracking, and pointing (ATP) systems. Fortunately, advanced optical and photonic systems for communication technologies are constantly being developed, which are very fast and extremely compact that will help in solving *SWaP* problems and will be absolutely needed for accurate and very fast switching to help in handling handover problems.

6.4.1 Satellite Constellation Networks

A single satellite is not adequate to deliver communication services to every location in the world since it can only cover a part of the world. For example, a communication satellite in geostationary orbit above the equator cannot see more than 30% of the Earth's surface [1]. The analysis is based on simple orbital geometry and geographical locations of various places in the world. In order to establish a complete coverage for communication, more than a single satellite is therefore needed such as a constellation of satellites. In a constellation of satellites, a number of similar types of satellites with similar capabilities move in the same orbit. It is also important to have a shared control of all the individual satellites. The object of this section is to introduce a constellation of satellites specifically designed for the purpose of establishing global communication. This section introduces satellite constellation, which will address three basic parameters (and their dependency to each other): (1) altitudes of satellites, i.e., the height of the orbit belonging to a given constellation; (2) number of satellites required with emphasis to the minimum number of satellites needed; and (3) number of optical ground stations (OGS) required for free-space laser/optical communications. Note that it is assumed that each single satellite will do similar function in a given orbit for a shared purpose.

Several commercial satellite constellation networks were designed and constructed in the 1990s, but later on many of them were not in practical operation. Some of the historical developments of constellation satellites include the following:

Low Earth orbit (LEO) *iridium* system, 66 active satellites constructed by Motorola that demonstrated the feasibility of Ka band (26.5–40 GHz) radio inter-satellite links (ISLs)

LEO *Globalstar* system (48 active satellites) based on CDMA-based frequency-sharing technology developed by Qualcomm

840 active satellites and 84 in-orbit spares in LEO orbits at 700 km altitude for broadband networking to fixed terminals: Ka band; 288 active satellites by Boeing (*Teledesic*)

80 satellites at 1400 km altitude (Alcatel's *Skybridge*)

10 active MEO (medium Earth orbit) repeater satellites by *ICO* (2002)

6.4.2   Constellation of Space Communication Satellites: Some General Description

Advances in optical technology and FSOC are bringing fundamental changes in the design of the global network infrastructure, which now includes constellation of optical satellites. Satellites in various constellations can beam data down to Earth using lasers to bring connectivity to places that either did not have any connectivity or had connectivity with very limited bandwidth. Data communication can be achieved at 100 Gbit/s between ground nodes and satellites in a constellation, whereas data can be exchanged at 200 Gbit/s between satellites. The ground laser light networks will include a number of efficient software-defined wide area networks (SD WANs) and will require the high-capacity advanced software to manage the global network. Laser beams are severely affected by the atmospheric effects such as turbulence and presence of clouds in the line-of-sight path. The network operating system therefore needs to be built that will be able to route traffic. Figure 6.9 illustrates an artist's concept of constellation of laser satellites interconnected with ground nodes and with each other to establish communication connectivity in a number of distant locations on the Earth which can include remote locations (drawn not to scale: for illustration purpose only). The optimum choice of a laser communication link between a satellite and a ground station can be selected to move around clouds, which interferes in beaming data between Earth's surface and its orbit. A machine learning model trained over several years with weather data will be used to predict the best routes automatically.

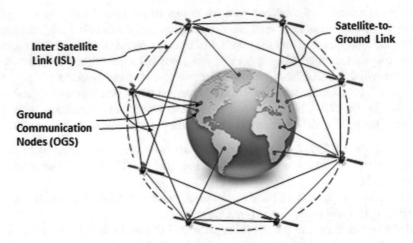

Fig. 6.9 An artist's concept of constellation of laser satellites interconnected with ground nodes and with each other. (Drawn not to scale: for illustration purpose only)

6.4.3   How Optical Satellite Constellation Can Achieve Global Networking?

Basically, data from the receiving satellite can be transmitted back to a ground gateway station, which then can be routed to the final destination via terrestrial infrastructure. In case of a satellite constellation, received data can be routed through the space network either via multilayer orbit satellites or via inter-satellite links (ISLs) to the last satellite that covers the area where the destination user is located in another geographical location. FSO communication links based on laser links offer the high data rates (~multi-Gbit/s or even more in future) required today because of the increasing traffic demand of the Internet devices and applications users. The other advantages of SWaP of laser transceiver components and subsystems suitable for using on the laser satellite constellation have already been discussed elsewhere. Another important consideration for justifying the satellite constellation for global communications is that there is also a potential to the deployment of a global Earth-orbiting wireless optical *backbone network*, which can also act as Internet satellites so that high-speed Internet connectivity can be easily obtained from *anywhere, anytime*.

6.4.4   How Optical Satellite Constellation (OSC) Can Be Designed with LEO Satellites?

OSC architecture for data combines the best of communications: ultrahigh speed as obtained with optical fiber and ubiquity of satellite. Existing satellite systems cannot satisfy today's connectivity requirements of increasingly data-driven and cloud-based business worldwide. Optical links between satellites at various orbits as well as ISLs can offer the advantages of communication such as latency, true global coverage, security, and flexibility in network configuration for point to point to multipoint data solutions. LEO satellite constellation is one of the first choices of designing OSC, but it does not have to be limited to LEO OSC since other OSC-using satellites in various orbits such as small satellite, CubeSat, and nanosatellite constellations can also be considered in the future for other types of architectures for high-speed data exchanges. For example, how tiny satellites could help warn of the next big hurricane with a constellation of nanosats and improve our understanding of the world's most dangerous storms has been recently reported [12]. A first-of-its-kind nanosatellite project is expected to launch its first qualification unit from Cape Canaveral aboard a Falcon 9 rocket. The constellation will consist of seven small satellites that will monitor Earth's tropical zone, which spans about 40° of latitude to the north and south of the equator. The nanosatellites must be launched into a very particular orbital configuration. A total of three pairs of these satellites will share an orbit at a slight angle to the equator—30°—on the opposite side of the globe from one another but following the same trajectory so that they will crisscross the equator at different points. It is interesting to note that the nanosatellites can also be equipped with optical communication modules so that high-data-rate FSO communications can be established with nanosatellite constellation at much lower orbital altitudes and with the possible least latency.

6.5   High-Level Design of Optical Satellite Constellation

Unlike a single satellite, a satellite constellation can provide global coverage such that at any time everywhere on Earth at least one satellite is visible so that a line-of-sight (LOS) communication link can be established.

The optical network design is based on the topology of the constellation of satellites and can provide connections with the minimum possible number of light-path hops, and the capacity of each light path can be determined by the number of connections sharing it. For inter-satellite links (ISLs), failure restoration can be provided through capacity reservation in the existing service light path using the redundancy inherent in the networks. The space network with the constellation satellites needs to be of very high throughput so that the space network can be connected seamlessly with the terrestrial backbone network.

A number of companies are placing and have future plans for placing a larger number of satellites in LEO. During the last 2 years, the number of active and defunct satellites in LEO has increased by over 50% to 5000, some of which are SpaceX to add 11,000 more in the building of its Starlink mega-constellation with permission filed with FCC for another 30,000 satellites; the other companies like OneWeb, Amazon, and GW (a Chinese state-owned company) have similar plans. This clearly shows that a tremendous number of communication satellites in very large constellations need to develop and design efficient communication architectures for routing, proper topology for onboard processing, and other communication schemes to handle data information exchanges accurately.

Basic concepts of design of constellation of communication satellites

Optical satellite constellations have the potential for providing customers with symmetric, very-high-speed, low-latency, and highly secure communications between locations anywhere on Earth. The laser/optical satellites can belong to LEO constellations or any other constellations of satellites such as small, micro, or nanosatellites at lower orbital altitudes to provide multi-Gbit/s data speed with very low latency. With the recent rapid development of satellite technologies and satellite-launching vehicle, LEO mega-constellation can also provide high-performance Internet access with global coverage.

6.5.1   Methodology of Defining a Satellite Constellation's Parameters Relevant to Satellite Laser Communication

There is a number of excellent books available in the literature on the topic of satellite technology specially addressing communication satellites, and therefore it is not repeated in this book. Most of the discussions are based on wireless communications with RF signals. Since the early 1990s, dozens of communication satellite constellation have been proposed and several have been designed to operate in space. A constellation of satellites includes a number of satellites in a given orbit belonging to that constellation. In order to develop the communication technology with a satellite constellation, it is important to understand the methodology of defining a satellite's basic parameters which include satellite's geometry and communication links between every two satellites via inter-satellite links (ISLs). The basic geometry and the appropriate links are useful in designing and developing satellites to provide communication and Internet access to the entire globe even to remote locations.

Each satellite's orbit can be described by the coordinates called Keplerian elements from which a satellite's location at any time can be calculated according to the following equation:

$$\text{Mean motion} \quad n = \frac{2\pi}{P} \tag{6.1}$$

where P is the orbital period which can be calculated from the Newton's form of Kepler's third law: "the square of the orbital period of a satellite is directly proportional to the cube of the semimajor axis of its orbit":

$$P = 2\pi \sqrt{\frac{a^3}{G(M_e + m)}} \tag{6.2}$$

where G is the gravitational constant ($G = 6.67408 \times 10^{-11}$ m³/kg/s²), M_e is the mass of the Earth ($M_e = 5.972 \times 10^{24}$ kg), and m is the mass of the satellite which can be ignored. Figure 6.10 depicts the position of a satellite in an orbit with respect to mean anomaly, the true anomaly, and eccentricity anomaly.

The position of an orbiting body in an elliptical orbit is given by the eccentric anomaly, E. The mean anomaly, true anomaly, and eccentric anomaly at an instant time t can be written as

$$M = nt = E - e \sin E \tag{6.3}$$

$$v = 2 \tan^{-1} \sqrt{\frac{1+e}{1-e}} \tan^2 \frac{E}{2} \tag{6.4}$$

$$d = a(1 - e \cos E) \tag{6.5}$$

LEO constellations usually have circular orbits where $e = 0$, and $M = E = v$, and therefore a satellite's location can be described by true anomaly only within its

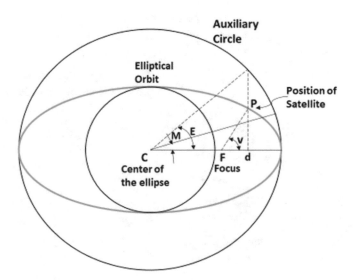

Fig. 6.10 The position of a satellite in an orbit with respect to mean anomaly M, the true anomaly v, and eccentricity anomaly E. (Drawn not to scale)

orbit. For any given LEO constellation (for example iridium constellation), if the satellite orbital elements are specified, the position of a satellite can be predicted to develop communication scenario and finally to evaluate communication system performance.

6.5.2   Various Satellite Earth Orbits

A quick overview of various Earth orbits is shown in Fig. 6.11 so that satellite communication systems can be designed for different satellites moving in various orbits with different altitudes of satellite constellations. Geostationary Earth orbit (GEO), medium Earth orbit (MEO), and low Earth orbit (LEO) together with Molniya orbit (Russian television) are shown. For elliptical orbits (generally at an inclination of 63.4°), the coverage is only provided when the satellite is moving very slowly relative to the ground, while at apogee, furthest from the Earth's surface, power requirement in link budgets is calculated for this distance. At low perigee, the satellite does not provide service coverage.

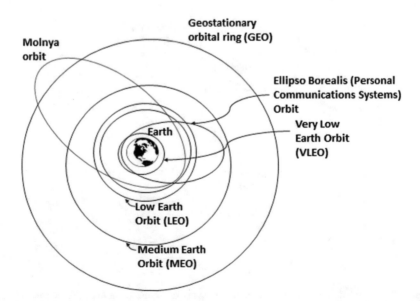

Fig. 6.11 Various satellite Earth orbits. (Drawn not to scale)

6.5.2.1   Constellation Design

A constellation of satellites and ground stations for communications access using high data transfer are needed for the data uploading and downloading, which are accomplished by the transmission and receiving satellites and ground stations. In designing constellations of LEO satellites to establish global coverage for communication between the satellites in the constellation and Earth stations, there are two main requirements to be considered:

1. It is absolutely required to maintain constant communication within the satellite constellation to be able to transfer all the received data to be transferred to any satellite using laser communication. All the satellites in a given orbital plane will therefore be placed in a "string of pearls" formation so that any satellite at any instant can exchange data to the adjacent satellite either ahead or behind. Laser communication system parameters such as power, transmitting beam divergence, receiving FOV, complexity of the ATP, and link range equation will then finally determine the *total number of satellites* required in each orbital plane.
2. Constant communication with 100% reliability will require a large number of satellites (over 80) and a large number of ground stations. To develop and design a larger number of optical ground stations (OGS) will depend on a number of factors such as cost, availability of geographical locations on the Earth (located in high altitudes, and mostly "cloud free"), sophisticated transmitting and receiving systems with complex adaptive optics (AO) capability to mitigate atmospheric turbulence between satellite and atmospheric path near the ground station, complex ATP system, as well as available terrestrial gateways to distribute data to users within the coverage area of a line-of-sight satellite. For example, a Walker constellation will require a large number of orbital planes with increased launch and operation costs.

With the advent of advanced laser communication technologies, it is becoming more and more feasible not only to provide broadband global Internet connectivity as discussed in the author's book [2] but also to minimize the number of satellites required. In addition, simply by allowing communications using many simple forwarding ground stations within the system, the fewer orbital planes will be needed within the constellation. Note that a detailed design of various LEO constellations is not addressed in this book because the main emphasis of this book is to develop advanced communication technology concepts and feasibility for laser communication satellite constellations to achieve the high data rate (multi-Gbit/s or much more) with space optical networks and terrestrial links, where the users can be located anywhere on the Earth including extreme remote places. Some general outlines for constellation designs will be discussed in this book.

Authors in a recent paper discussed a trade-off analysis with simulations for different constellation designs, number of satellites, satellite altitudes (400–500 km), satellite distances (2000–5000 km), and communication time gaps in the data transfer [14]. The number of satellites required in each orbital plane for a constellation at an altitude of 500 km with a 5000 km distance between each satellite and the next is given by [14]

$$\frac{2p(R_e + 500)}{5000} = 8.6 \tag{6.6}$$

where R_e is the radius of the Earth = 6378 km and p is the number of equally spaced plane according to Walker's notation. Satellite numbers are integers, and therefore 9 satellites are required within each orbital plane, and for three orbital planes the minimum number of satellites for the constellation required is 27. The number of ground stations can also be considered as ground constellations, and therefore possible optical wireless communication scenarios and analysis need to be done for the complex loops combining these two space and ground constellation networks. Communication access is considered whenever there is any satellite within the constellation that has a line-of-sight (LOS) access to any satellite ground station. The number of potential ground stations in various parts of the countries on the Earth can be a parameter in the communication analysis to develop reliable and efficient global laser communication systems. In the satellite constellation design, atmospheric effects like atmospheric turbulence and scattering medium need to be considered in the bit error rate (BER), probability of fading, and end-to-end link analysis for a highly reliable secure communication. The maximum angle relative to the zenith for a satellite transceiver to point toward the receiver to transmit data is another important parameter in the constellation design because it also affects the communication performance due to the atmospheric effects in determining the BER. Clouds, fog, and rain, which can limit the laser propagation through the atmosphere, are another important consideration in which redundant transmissions to other Earth terminals in different location will be required to avoid interruptions in communications. Future development of fast, portable advanced laser/optical devices and modules will allow to help in laser satellite constellation design with reduced number of satellites as well as smaller number of ground stations by adding mobile ground stations.

6.5.3 Constellation Design for LEO Mega-Constellations for Integrated Satellite and Terrestrial Networks

An efficient integrated network of satellite network and terrestrial network sharing the same priority is required when LEO mega-constellation (i.e., with 100s or even 1000s of satellites) is considered. As mentioned earlier, laser communication satellites have the great potential of providing high-speed communication and Internet access to the entire globe, especially to remote locations. Since 1990s, dozens of communication satellite constellations have been proposed for LEO, MEO, and GEO satellites. Lower altitude satellites ensure lower latency for better network performance and quality of service (QoS), while large satellite number can provide higher network capacity with added advantage of network stability using extra redundancy. With increased number of next-generation LEO satellite mega-constellations, the difficulty for network management and packet routing process is dramatically increased. The most optimal route is therefore very difficult because satellites travel with high dynamic situations.

Inter-satellite links (ISLs) for the constellations. ISLs can be between two satellites on the same orbit plane and also between different orbit planes. Figure 6.12 shows an example of inter-satellite distances for both nearest and farthest situations in a laser satellite constellation.

The distance P_1P_2 between the two adjacent satellites in the same orbit can be found from the equation

$$d_s = \frac{R \times \sin \alpha}{\sin \phi} \tag{6.7}$$

where

$R = R_E + r_s$, R_E = the radius of the Earth (6378.14 km), and r_s = orbit height of the satellite above the Earth

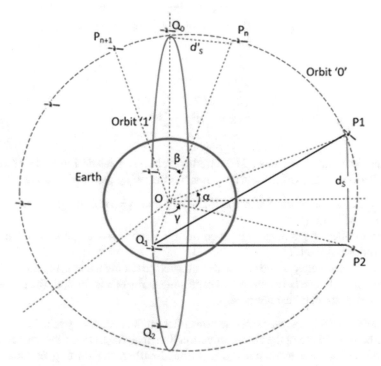

Fig. 6.12 An example of inter-satellite distances for both the nearest and farthest situations in a laser satellite constellation

α = angle subtended by the two adjacent satellites = $\dfrac{360°}{n_S}$

n_S = number of satellites in a given orbit ≠ the total number of satellites in the constellation

$\phi = (180° - \alpha)/2$

For interplane ISL satellites in nearby orbits, the nearest situation occurs when a satellite in an orbit flies over the north pole or south pole.

In this case, for two nearby orbits for the "nearest case," $\phi' = (180° - \beta)/2$, where $\beta = \alpha/2$, so that the distance Q_0 between the nearby orbital satellites is given by [13]

$$d'_s = \frac{R \times \sin \beta}{\sin \phi'} = \frac{R \times \sin(\alpha/2)}{\sin\left[\dfrac{(180° - \beta)}{2}\right]} \quad (\text{nearest case}) \tag{6.8}$$

The distance for the "farthest case" $Q_1 P_1$ between the satellite at P_1 of orbit "0" and the other satellite at Q_1 of the nearest orbit "1" can be found from the equation [13]

$$Q_1 P1 = \sqrt{\left(\frac{d_s}{2}\right)^2 + R^2 + a^2 - 2R \cdot a \cdot \cos\gamma} \tag{6.9}$$

where

$$a = \sqrt{R^2 - \left(\frac{d_s}{2}\right)^2} \tag{6.10}$$

Therefore, the above sets of equations can be used for designing laser satellite constellation system with ISLs if one has the following parameter values:

- R_E = the radius of the Earth (6378.14 km) and r_s = orbit height of the satellite above the Earth (km).
- n_S = number of satellites in a given orbit; calculate α = angle subtended by the two adjacent satellites.
- The distance between the two adjacent satellites in the same orbit can be found.
- Also, the distance between two satellites in nearby orbits for both the nearest and farthest cases can be computed.

In order to design a laser satellite constellation, it is necessary to select the satellites to be located along the communication line of sight (LOS) at the source as well as at the destination locations. As mentioned earlier, the starting parameters for designing the constellation are the number of satellites (and their locations) in the constellation and the availability of the optical ground stations in order to maintain uninterrupted communications to anywhere on the Earth, especially to remote areas.

6.5.4   Optical Inter-satellite Links (ISLs): Implementing a High-Speed Communication Link Between Satellites

In the previous section, constellation design is addressed with specific emphasis on determining distances between satellites in a constellation. Distances between the adjacent satellites and also the farthest distance between the satellites in nearby orbits were discussed. Accurate values of these distances are necessary to evaluate ISL performance for providing efficient and reliable satellite communications by integrated space/terrestrial networks. Space-based optical communications using satellites in LEO constellation have the potential for the proposed Internet in the Sky network of the future. Laser communications can provide high-performance inter-satellite links. Some of the basic requirements of the space-based lasers and optics used for beam forming and also receiver optical gain and detectors used in FSO communications are important since they are critical in designing successful inter-satellite communications system. Orbiting satellites LEO-LEO or GEO-GEO cross-link the constellation by communicating with each other and finally have the capability to beam high-speed data to ground stations. Some of the parameters necessary to design and evaluate the communications system performance on board a satellite are size, weight, and power and the cost (SWaP-C). Laser links around 1.55 µm region (~200 THz) can have bandwidths even higher using multiplexing scheme (WDM). Advanced WDM technology and tunable wavelength lasers allow to increase the bandwidth with four or more different wavelength lasers. Optical system with coherent beam has very narrow beam width, which can be focused to extremely small spot sizes approaching diffraction limit. For LEO constellations, a narrow laser beam width will have a great advantage of encountering very small interference to and from adjacent satellites. Because of the extremely small narrow laser-transmitting beams, much more accurate acquisition, pointing, and tracking (APT) system is required specially for the spacecraft moving at 3 km/s for GEO to 7 km/s for LEO satellites. This is technologically very challenging using existing optical and photonic devices and modules on board. Very small aperture sizes are required for optical antennas, for example ~15 cm aperture to operate at about 1 Gbit/s. Considerable size and weight savings in the satellite payload for laser satellite constellation are therefore possible.

6.5.4.1   Key Challenges for Implementing a High-Speed Optical ISL Between Satellites

Pointing accuracy and acquisition of the link: For laser sources the typical beam width and with a few cm size apertures, the beam width is a few μrad and the pointing accuracies in acquisition of the link, and zeroing and tracking other spacecraft is an extremely difficult problem. In addition, enormous speeds of the satellites in the constellation including the jitter and vibration of each spacecraft introduce enormous challenges to design a stable optical ISL link. Recent advances in the photonic and optical technologies such as MEMS-based agile beam steering and adaptive optics (AO) and micro/nano-optics and developments of advanced optical chips can help in the precision pointing and acquisition in space. Furthermore, vibration and shaking at around 400–1000 Hz range experienced by satellites in orbit need to be compensated for accurate tracking. Background noise from the Sun and bright sources needs to be considered since it contributes to the receiver noise, which affects the available communication data rates. In short, careful designs of laser/optical transmitters, low-noise optical receivers, advanced APT system, and other subsystems are necessary for designing successful laser constellation communication systems.

6.6   Optical Ground Stations for Global Coverage Using Laser Satellite Constellations

As discussed in earlier sections, laser light communications have the potential for deploying an all-optical global communications network to provide connectivity to remote and formerly unreachable locations worldwide. The laser satellite constellation of LEO satellites will have the potential service capacity in excess of Tbit/s with multi-Gbit/s optical cross-links (ISL) and satellite/terrestrial optical bidirectional links. Optical satellite backbones will be able to interconnect a tremendous number of presences clustered within a number of software-driven WANs around the world. Some of the advantages of laser/optical satellite networks include flexible, very fast intelligent routing and routes; flexible points of origination/destination; capability of very-high-speed data exchanges to satisfy today's demand for continuous increasing data; and connectivity requirements. Secure communication is also an important consideration in global broadband Internet connectivity. One of the most important challenges for successful deployment of laser technology is atmospheric effects such as atmospheric turbulence and cloud coverage along the line-of-sight propagation link between a satellite and an optical ground station (OGS). In order to mitigate this problem, it is proposed to build a network of interconnected and geographically diverse OGSs located in various geographical locations on the Earth so that at any time at least one space-to-ground optical link is available to connect one satellite in the constellation. Researchers in a recent paper

[16] presented a new approach for quantifying the availability of an optical ground network based on cloud fraction data to estimate the probability of having a number of link fails due to cloud coverage.

If the cloud coverage interrupts the downlink laser from a spacecraft to a ground station, another ground station is automatically chosen by the spacecraft. The link outage probability (LOP) and the optical ground network availability (ONA) are given by [16]

$$LOP = f_\chi (X = N) = \mathcal{P}(\chi = N),\tag{6.11}$$

$$ONA = 1 - LOP\tag{6.12}$$

where $f_\chi(X)$ is the probability of having a given number of space-to-ground link fail due to cloud coverage, N is the number of ground stations, and χ is the sum of channel model random variable χ_i assumed to be distributed according to Bernoulli distribution: $\chi_i \sim B(p_i) = f(x) = \begin{cases} 1 \\ 0 \end{cases}$

$$\chi = \sum_{i=1}^{N} \chi_i\tag{6.13}$$

The authors discussed a framework for approximating the availability of an optical ground framework based on an ON/OFF channel characterized by the probability of having a disrupting cloud. The probability of having a cloud in the link for any ground station can be estimated from cloud modeling providing cloud fraction data product information. Ground station spatial and temporal correlations (across ground stations) are also addressed in their paper and recommended to use the monthly averaged cloud fraction dataset at the ground station and to obtain $f_\chi(X)$ for different correlations and various number of ground stations.

A model to optimally determine the location of optical ground stations (OGSs) for LEO satellites is reported [17], which can take into account correlated cloud coverage among all visible ground stations. Both spatially isolated ground station model and the two spatially correlated ground stations are considered in their work. The model considers trade-offs between minimal cloud probability, minimal latency, and proximity to supporting infrastructure in selecting locations for networks of ground stations and thus presents a scheme to optimally determine the location of OGS for LEO satellites. Both the best existing stations and the ground stations of unconstrained scenarios located at any point on the Earth surface are considered in their paper. Free-space optical (FSO) communication provides the high-speed communications (10s–100s of Gbit/s) between LEO satellites and OGSs. The results show that the main parameter in selecting the optimum location is the mean cloud probability over each OGS, which results in locations in the ±20° to ±40° latitude band to be the most appropriate candidate sites instead of traditionally polar latitude sites.

In another study of optical ground station reported by DLR [18], some of the OGS technical status is reviewed. One of the OGSs has been constructed on the Canary Islands as part of the SILEX program and will be used for commissioning and routine checkout of the ARTEMIS satellite as well as to receive and transmit data. The OGS is situated at an altitude well above the first inversion layer or cloud level. Link probabilities were predicted by analyzing cloud data on a global scale, which can be improved if data from meteorological uncorrelated sites from several stations can be used for predicting the actual statistical probability. Cloud correlation distance and radius of the strongly correlated area are approximately 500 km based on the analyses of two stations, one at Sardegna and the other in Egypt. For a network of several stations, the combined probability of meteorologically uncorrelated stations is given by [18]

$$P(n) = 1 - \prod_{i=1}^{n} P(C_i) \qquad (6.14)$$

where $P(C_i)$ is individual cloud coverage probability for ith station and n is the number of stations.

For example, for three stations with fully uncorrelated coverage S of 20%, the combined link probability would be 99.2%.

6.6.1   Satellites Supported by Various Relay Network Configurations

Relay communication satellite constellation has the potential of significantly reducing system response time. However, because of low-cost satellite and launcher technology, the design of relay communication satellite constellation is becoming now more practically viable. Communication platforms using low Earth orbit (LEO) and very low Earth orbit (VLEO) satellites can achieve persistent inter-relay inter-satellite links (ISLs) and thereby minimize system response time. The constant demands for the amount of information data at very high speed can thus be satisfied very well because of recent developments of communication technologies, one of which is by relay communication scheme using some of the following methods:

- Design relay satellite systems using geostationary Earth orbit (GEO) relay satellite systems for tracking and data relay
- In addition, design non-GEO relay satellite systems (MEO, LEO, VLEO) for tracking and data relaying

It is therefore important to develop communication tools and techniques that allow heterogeneous satellite constellations to provide simultaneous connectivity between LEO satellites and ground stations using relay satellites in MEO (or another LEO in another orbit). Different LEO satellites can be used which can be in different orbital heights. Improvement of the communication system performance as ISLs between the relay satellites can then be analyzed. Orbital mechanics can determine completely the prediction of inter-satellite and satellite-to-ground visibility for performing line-of-sight (LOS) optical communications. The cost can also be a constraint in designing a reliable and efficient design of relay network configurations. The total cost of the mission is composed of constellation cost and launch cost, where the constellation cost can be computed from multiplying the number of total satellites by the cost of a single satellite and the launch cost by multiplying the cost of a single satellite by the number of planes in the constellation.

For achieving laser satellite communication, the main problem is the satellite LOS with either GOSs or LEOs. For example, depending on the satellite altitude, LEO satellite completes one orbit every 90 min, and it requires a clear LOS from the optical ground station (OGS) to the LEO satellite. However, the window of visibility during a given pass may last only 10 min or so, and it will take many hours before the satellite flies over overhead again. One possible solution to overcome this problem is to install more OGSs around the world to increase the space-to-ground communication time windows. Relay satellite systems with laser communication satellites can be designed properly to relay user data in almost real time at very high data rate. Both GEO and non-GEO (such as LEO and MEO) relays can be properly designed. A large number of satellites in data relay network in LEO can fly in polar or inclined circular orbits so that global coverage can be achieved by coordinately spacing the satellites in such a way so that when a particular single satellite loses LOS there will be at least another satellite into view. The small orbital periods, for example from 90 to 130 min approximately, limit the contact windows between the relay satellites and the GEO satellites to about 10–15 min. Relay system based on LEO satellites will have very short satellite-to-satellite contact windows to perform the needed accurate acquisition, pointing, and tracking (APT) system. Recently, researchers have described relay constellation design [19] on an equatorial plane for MEO satellite constellation using bidirectional interlayer ISLs in the space network. Any constellation is characterized by altitudes, number of orbital planes and types, and number of satellites per plane. Note that relay network configurations use the relay satellites either in *bent-type* or in *store-and-forward* communication technology, which can include multiple-layer satellite networks as discussed earlier in this chapter. The types of ISLs in multiple-layer satellite networks can be *intra-orbit* ISL (links in same orbit and in same layer), *interorbit* ISL (links connecting satellites in adjacent orbits in the same layer), and *interlayer* ISL (connecting satellites in orbits at different layers). Therefore, relays can be achieved *with* and/or *without* inter-satellite links. Some of the parameters for intra-orbit ISLs are azimuth angle, elevation angle, and transmission distance between satellites. The intra-orbit ISL angle α between two adjacent satellites, Sat i and Sat j, is given by [19]

$$2\alpha = \pi - \left(\beta + \beta'\right) \tag{6.15}$$

where $(\beta + \beta')$ is the angle between two adjacent satellites and is given by

$$\left(\beta + \beta'\right) = 2\pi / N_{sp} \tag{6.16}$$

N_{sp} is the number of satellites per plane.

The angle α can therefore be written as [19]

$$\alpha = \pi/2 - \pi/N_{sp} \qquad (6.17)$$

Because of the requirement of LOS path between any two adjacent satellites to be above the altitude of R_{min}, the minimum shadowing radius of Earth $R_{min} = R_E$ (the Earth's radius) + L_{min} (atmosphere's altitude). The value of L_{min} is usually the Kármán line (border between the atmosphere and the outer space) and is assumed to be 100 km. No link between two adjacent satellites occurs when the Earth central angle ($\beta + \beta'$) is 180°. The only possible LOS ISLs are one link between one satellite S_1 and the second satellite S_2, second link between satellite S_2 and the third satellite S_3, and no ISL between S_3 and S_1. Therefore, the minimum number of satellites per plane $N_{sp} = 3$ to avoid Earth shadowing. The minimum value of α is also given by [19]

$$\min \alpha = \sin^{-1}\left(\frac{R_E + L_{min}}{R_E + h}\right) \qquad (6.18)$$

where h is the satellite orbital altitude. From the results of the author's research work [19], the minimum number of required satellites per plane increases as the orbital altitude is decreased; for example, for an altitude of about 6571 km, the minimum number required $N_{sp} = 3$, and for 570 km, the minimum number of satellites required is $N_{sp} = 9$. The distances between two adjacent satellites at the same N_{sp} are increased with increased orbital altitudes. For a valid constellation with continuous interorbit ISLs between all satellites, the minimum number of interorbit cross-links, $\min(N_{cross})$, can be written as

$$\min\left(N_{cross}\right) = N_{sp} * (P - 1) \qquad (6.19)$$

where P = number of planes for the relay constellations. The results for $P = 2$ and $N_{sp} = 4$, and the value of the minimum number of interorbit cross-links is $\min(N_{cross}) = 4$.

The altitude dependence of the distances between two adjacent satellites can be found from the equation [19]

$$D_{ij} = \sqrt{2}\left(R_E + h\right)\sqrt{1 - \cos\left(\frac{2\pi}{N_{sp}}\right)} \qquad (6.20)$$

where D_{ij} = the distance between two adjacent satellites with the same minimum number of satellites per plane, N_{sp}. When the satellite orbital altitude h is increased, the value of the distance between two adjacent satellites also increases for the same of N_{sp}.

Various satellite constellation types are being considered because of the development of recent communication technologies, which have led to cheaper launch costs. Recent developments address the design of large low Earth orbit (LEO) constellations to deliver high-throughput broadband services with low latency with plans of launching thousands of satellites by a number of companies like SpaceX, Amazon, and OneWeb. For example, SpaceX is developing to build its Starlink constellation with many thousand satellites. There has also been a medium Earth orbit (MEO) constellation of 20 satellites in a circular orbit along the equator at an altitude of 8063 km. A new type of constellation is also being designed and developed called "hybrid constellation," which combines assets in different orbits, for example between MEO and GEO, or backhauling of LEO satellite data through higher orbit satellites like MEO or GEO so that the terminals can seamlessly hand over data between two orbits for achieving uninterrupted connectivity. Another trend in designing a constellation of satellites is the recent development of *onboard processing capabilities* using FSO components to provide flexible and very fast routing/channelization, beam forming, and other communication tasks. *Nonterrestrial space networks* are being recently designed and developed to integrate satellites, high-altitude platforms (HAPs), and unmanned aerial vehicles (UAVs) as discussed earlier in the multilayer network context. The main objective is to seamlessly integrate the assets in various platforms to complement the terrestrial infrastructure in most efficient ways. In addition, cube/micro/nanosatellites are being developed which will also be integrated in the optical satellite constellation to allow easy access to space in terms of establishing global bidirectional communication connectivity from anywhere on the ground directly via satellites.

6.6.2 Laser Satellite Constellation Networks

Because of the recent advanced developments of high-speed FSO cross-link, a satellite laser communication network as a part of a larger integrated space/terrestrial network is now entirely possible. Multi-gigabit laser inter-satellite links are useful in routing traffic through the space segment serving as a global space-based backbone network. Inter-satellite links help to route communication data via space network up to the satellite that covers the destination area of the intended user's location. There is a potential for constructing a laser/optical satellite constellation which can consist of geosynchronous Earth orbit (GEO), medium Earth orbit (MEO), low Earth orbit (LEO), small/CubeSat/microsat, nanosat, high-altitude platform (HAP), and unmanned aerial vehicle (UAV) laser/optical equipped communication terminals. Multiple constellations working together can also significantly improve the performance of the optical space network. FSO communication in laser communication constellation networks can deliver very high data rates between satellites at 10s to even 100s of Gbit/s to satisfy the constantly increasing traffic demand. The backbone relay satellites will require multiple apertures where optics with high antenna gain and modest transmitter power can deliver needed optical power (from the link budget analysis) to a small area at the receiving satellite. For example, a link distance equal to one-time synchronous orbit (44,000 km) for optical wavelength of 1.5 μm with 1–20 W power may need only a few cm (<30 cm) aperture diameters. In addition, with FSO, it is also easy to apply multiplexing, demultiplexing, switching, and routing operations [7].

Latency and propagation delays are also important factors in designing a constellation satellite. Latency is the time delay over a communication link and is usually measured as a round-trip time in msec. Latency determines how quickly the user begins to get a response from the server. The higher the altitude of the satellite, the more the latency. Table 6.1 shows latency values for various satellites at different altitudes:

6.6.3   Satellite Constellation Network Physical Topology

The main purpose of developing a constellation of space communication network is to establish communication link and connectivity worldwide *anywhere, anytime*. A user from any location should be able to communicate with any other user located anywhere in the world. The communication also needs to be bidirectional (transmit/receive) and uninterrupted. Because of the demands of increasing data requirements at high speed, optical communication is a potential viable solution. The physical topology of the space backbone satellite constellation is an important consideration so that a backbone constellation can provide the coverage required by the users located on the grounds worldwide. There can be a number of different constellations with different altitudes, number of orbital planes, and arrangements of satellites within the orbits and also a number of required ground gateway stations. A concept of pass-through traffic which increases cross-link capacities in a satellite constellation is discussed [7]. The ratio of pass-through traffic to add/drop local traffic, for example for a LEO or MEO constellation, in their coverage area is given by $\sim(m + n)/4 - 1$, where m is the number of planes and n is the number of satellites per plane with the configuration that all the satellites are connected to their neighbors [7]. Optical links have the potential for the backbone to provide about 100 Gbit/s communication capacity. An ideal satellite constellation configuration design should

Table 6.1 Latency values for various satellites

Configuration	Altitude (km)	Latency (ms)
GEO	~36,000	240
MEO	2000–20,000	13–130
LEO	200–2000	1.3–13
VLEO/CubeSat	100–450/500–900	0.67–3/3.3–6
HAP	15–25	0.1–0.17
LAP	0–4	0–0.027

achieve the desired ground coverage with a desired ground terminal minimal eleva-tion angle using the minimum possible number of satellites. The number of satellite nodes N placed in a grid configuration is given by [8, 9]

$$N = m \cdot n \tag{6.21}$$

where n = number of orbital rings and m = number of satellites per orbit. As a simple example, the total number of satellites for various configurations is the following [7]:

- GEO (altitude 35,786 km): $m = 3, n = 1, N = 3$
- Polar LEO space (altitude 1550 km): $m = 4, n = 3, N = 12$
- Polar LEO Earth (altitude 1550 km): $m = 8, n = 5, N = 40$
- Walker LEO space (altitude 1550 km): $m = 2, n = 5, N = 10$
- Polar MEO space (15,000 km): $m = 3, n = 2, N = 6$
- Polar MEO Earth (15,000 km): $m = 4, n = 2, N = 8$
- Walker MEO (15,000 km): $m = 1, n = 5, N = 5$

Since the satellites are moving continuously, the network traffic demand also changes affecting the complex dynamic connection for the traffic matrix between the source and the destination satellites. The average distance between the network pair (average connection lengths) is given by [10]

$$h_{av} = 1/(N-1) \sum_{h=1}^{h=d} h \, N_h \tag{6.22}$$

where

N_h = the number of nodes from a given starting node with h inter-satellite links (ISLs)
d = network diameter

The average number of satellites traversed by a connection is

$$h_{n\text{-}av} = h_{av-1}$$

The total number of connections in a fully connected network is $L = N \cdot (N-1)/2$, and the average number of connections transiting a node can be expressed as [10]

$$N_{TR} = L \cdot h_{n\text{-}av} / N = \left[N \cdot (N-1) \cdot h_{n\text{-}av} \right] / (2N)$$
$$= (N-1) \cdot h_{n\text{-}av} / 2 \tag{6.23}$$

For 2π constellations, the actual number will depend on the nature of traffic rout-ing, and in π constellations it will depend on the position of the node in the network topology. Each node adds/drops $N_{AD} = N - 1$ optical connections, and therefore the ratio of transit and add/drop can be written as [10]

$$r = \frac{N_{TR}}{N_{AD}} = h_{n\text{-}av} / 2 \tag{6.24}$$

The other issues in network traffic demand include (1) end-node pair connection bandwidth and (2) onboard processor load, which is the total amount of traffic handled by the onboard processor (OBP), P_{OBP}, and is the sum of the uplink/downlink capacity $C_{U/D}$ and the bidirectional transit capacity C_{TR}. P_{OBP} is given by [10]

$$P_{OBP} = C_{U/D} + C_{TR}$$
$$= C_{U/D} \left(1 + c \cdot \frac{N_{TR}}{N_{L\text{-}D}} \right) \tag{6.25}$$

where c = percentage of $C_{U/D}$ generating the long-distance traffic. The second term in the bracket, namely $c \cdot \dfrac{N_{TR}}{N_{L\text{-}D}}$, is the capacity of a long-distance connection. The details of the network traffic demand can be found in the paper by Nikos Karafolas and Stefano Baroni as cited in Ref. [10]. These parameters are important in designing the satellite constellation networks.

Optical satellite constellation where each satellite is equipped with advanced laser transceivers, routers, ultraprecise pointing, acquisition, and tracking (PAT) systems, high-power lasers, efficient doped fiber amplifiers, ultrasensitive photon detectors, etc. will have the most viable potential solution to provide broadband global connectivity to anywhere worldwide. At the same time, broadband satellite Internet applications to satisfy the demands of tremendous data traffic are increasing where the constellation of satellites in various altitudes and orbits is becoming essential. Data can also be routed through space network using very-high-speed (at 10s of Gbit/s to 100s of Gbit/s) inter-satellite links (ISLs) from ground gateway station to the destination user, which is covered by the last satellite of the space satellite network. Some of the critical issues to develop integrated *all-optical* satellite/ airborne and terrestrial LAN with a complete space optical satellite network are discussed here. There are two main configurations applied for the constellation of satellites: (1) π constellations (or Walker "star" pattern), satellites ascending from one hemisphere (180°), and (2) 2π constellations (or Walker "delta" pattern), satellites ascending from both hemispheres (360°). Note that a laser satellite in a constellation will need to process multi-gigabit traffic above hotspots on Earth at a given instant but may move to another location in space with no available laser links with hotspots. It is therefore essential to adapt with constant connection traffic matrix changes for maintaining different coverage at different geographical locations of both the source and the destination optical satellites. In the future, optical satellite networks will require the uplink communication capacity of 10 Gbit/s or more so that ISL needs to provide communication bandwidth in 10s of Gbit/s or even more. With laser inter-satellite links for high-speed data exchanges, it is possible to handle each source-to-destination node-pair connection with advanced communication technology. With the developments of sophisticated advanced wavelength division multiplexing (WDM) and demultiplexing technology schemes, the satellite uplink capacities can be greatly increased much beyond 10 Gbit/s when an optical transport network (OTN) across the optical constellation is designed. In addition, both single-hop and multihop optical connectivities need to be designed in the space optical network architecture. For developing the technology concept of optical satellite constellation, either a single or a multihop optical/laser connection can serve each node pair to determine the minimum number of satellite pairs needed at the optimum orbit. The selection of the matrix approach for designing the satellite constellation will of course depend on "*handover*" for routing. Note that for a large number of satellites in an optical satellite constellation, *multihop optical connectivity* will be required.

All-optical concept technology for establishing optical wireless communications for broadband global communications and Internet connectivity to *anywhere, anytime* is feasible using the recent fast and portable optical photonic devices, lasers, detectors, and recent wavelength division multiplexing schemes. Optical satellite constellation integrating satellite, aerial, HAP, UAV, and terrestrial terminals can be integrated to design space optical networks to establish communications and broadband global Internet connectivity.

6.6.4   Laser Communication Networks for Satellite Constellations

Space-based optical communications using constellations of satellites in low Earth orbit (LEO) and geosynchronous orbits (GEO) offer to develop high-data-rate (multi-gigabit) space/Earth links and finally to design Internet in the Sky network of the future. In the constellations of LEO-LEO or GEO-GEO, orbiting satellites can communicate with each other as well as beam data to ground stations using crosslinks in the constellation. Note also that an accurate acquisition, tracking, and pointing (ATP) system is needed for the spacecraft moving at 3 km/s for GEO and 7 km/s for LEO satellites. Space laser communications system performance will require critical designs of small optical antenna size, narrow beam divergence, and narrow field of view for space-based lasers and optics. Figure 6.13 shows an example of possible link configurations to establish communications from a sender user at a location to a destination user at another geographical location using both ground-to-satellite and inter-satellite links using satellites in a constellation group. It is also possible to design other hybrid link configurations using both interorbital and inter-satellite links. Each terminal must have laser transceivers to perform two-way (duplex) communication at high data rates (in excess of 10 Gbit/s and multi-Gbit/s) with high communication capacity. It is therefore obvious that integrated satellite design in a constellation is very challenging specially when the links also include terrestrial and have to operate in the presence of atmospheric turbulence and scattering medium. The advanced communication technology developments in laser and

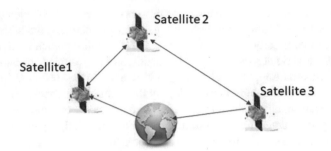

Fig. 6.13 Possible laser links from configuration in communication satellites in a constellation to establish communication from a sending source to a remote destination using point-point and cross-link links

photonic technologies in transmitters (source), receivers (detectors), and optical telescopes using adaptive optics (AO) and MEM-based agile beam steering, micro-optics, high-speed small-size silicon photonic chips, advanced wavelength division multiplexing systems, and sophisticated optical modulation techniques are just to name a few. Some of the basic characteristics of these integrated space/terrestrial network using satellite laser communication network in various constellations are discussed. The performance of the communication network can be drastically improved if multiple satellite constellations are used which can increase the capacity, reliability, and performance in global scale communication. Note that the number of satellites in a layer decreases with the increasing altitude so that the number of GEO satellites is smaller than the number of MEO satellites, which is also smaller than the number of LEO satellites. Over thousands of satellites are being proposed to be installed in the LEO layer.

Figure 6.14 shows a hierarchical organization for global optical space network where the LEO constellation is divided into grid points placed with equal angular distances. Routing information from the predecessor satellite is used by the successor satellite to maintain a logical location of switching to a new LEO satellite. As shown in Fig. 6.14, the management of various groups is assigned to a satellite in the upper layer. Users within the same footprint cell are covered by the LEO satellite directly in the line-of-sight (LOS) link. Users within a footprint cell can thus send information to that LEO satellite, which then routes eventually to the final LEO satellite directly above the LOS to the user in the destination location. In this way, communication information from a sender's location can be received by the user in any other geographical location. The concept of using LEO satellites to establish two-way FSO communication can be extended to include another topology of optical satellite constellation in lower altitude orbits with small satellites or CubeSats forming another layer of constellation.

Figure 6.15 shows the integrated space optical networks connecting GEO and LEO constellation served by terrestrial gateway (OGS) and HAP, UAV, and aircraft terminals in multilayer space networks. Terrestrial communication terminals can be directly connected to satellites in the constellation.

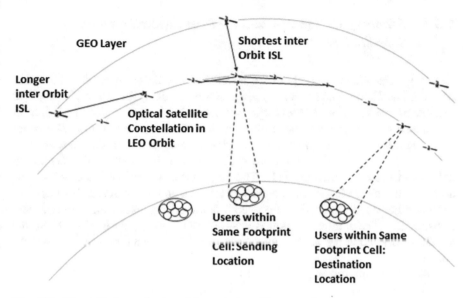

Fig. 6.14 Hierarchical organization of the optical satellite network

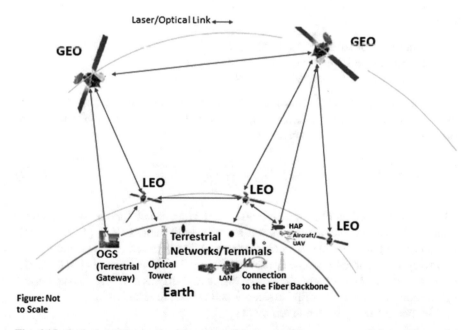

Fig. 6.15 Optical satellite routing between multilayer satellites, over cross-links and terrestrial network—an architecture of space/Earth integrated high-speed laser/optical communication network. (Figure: Not to scale)

6.6.5   Topology Design Method for Multilayered Optical Satellite Networks

Recently, researchers have reported a novel topology design method to minimize both link switch time and network average end-to-end delay for multilayered optical satellite networks. A reliable and efficient network is essential since laser link in satellite network requires higher link reliability and longer setup time. The paper proposed a topology design strategy based on non-equal-length snapshot division with the joint consideration of time delay (linking switching delay and transmission delay) and network reliability [11]. The proposed topology also ensures the reliability of the network for fully connected network and to maximize the degree of each satellite node minimizing the overall delay arising from link switching and transmission. The authors developed near-optimal topology design to achieve the smallest possible delay at the same time satisfying reliability and connectivity constraints.

The link switch time is given by [11]

$$t_{\text{link\_switch}} = 2\left(t_{\text{S}} + t_{\text{L}}\right) + t_{\text{r}} \tag{6.26}$$

where

t_{S} = the duration of adjusting angle of view
t_{L} = the path transmission time
t_{r} = the conversion time between the laser and the detector

The authors used as an example for the farthest communication link distance of 70,000 km the value of $t_{\text{link\_switch}} = 2$ s, otherwise $t_{\text{link\_switch}} = 0$ s for which average link switching delay can be written as

$$T_{\text{link}} = \frac{1}{M} \sum_{i=1}^{m} t_{i,j} \tag{6.27}$$

where M is the total number of links and $t_{i,j}$ is the link switch time between v_i and v_j that are the components of the vertex set of satellite nodes, $V = \{1, 2, \ldots, M\}$. Denoting a binary variable a_{ij} to represent visibility between any two satellites so that when v_i is visible to v_j, $a_{ij} = 1$; otherwise, $a_{ij} = 0$ when v_i is not visible to v_j. The paper also defined the symmetric $M \times M$ distance matrix as d_{ij} as the distance v_i and v_j. The connection relationship between v_i and v_j is represented as h_{ij}.

The average delay is then given by [11]

$$T_i = \frac{1}{M} \sum_{i=1}^{m} t_{ij} + \sum_{l=1}^{m} \frac{f_l}{C_l - f_l} / \sum_{i=1}^{W} \lambda_i \tag{6.28}$$

In the above equation, m is the number of links in the network, W represents the total number of businesses assuming the arrival of each service to follow a Poisson distribution, C_l is the lth link capacity, and f_l is the load of the lth link. Note that the upper bound of the laser channel capacity can be estimated using Shannon's formula for communication n theory in presence noise. Applying the degree constraint d_i to be equal to the number of laser terminals on satellite v_i, the authors developed the topological design problem with two minimization objectives related to average transmission delay and link switch delay:

$$\min \frac{1}{T}\left(\sum_{i=1}^{k}\left(T_{\text{link}}^{(k)}+T_{\text{trans}}^{(k)}\right)\right) \tag{6.29}$$

where k is the number of snapshots (referring to a period when the topology connection is fixed) and T is the network period. Minimum delay topology design (MDTD) is discussed in their paper with an algorithm and simulation. Their proposed research described an effective strategy to determine the relationship between the joint consideration of snapshot division and network delay and developed an MDTD algorithm to study the trade-off between link switch and link availability. Comparing the two parameters such as the number of link switches and the corresponding network average delay, their results also show that the proposed MDTD is more efficient than the maximum visibility topology design (MVTD) in which case one satellite is always connected to other satellites with the longest visible time. Design of laser satellite constellations for various orbits for establishing communications and Internet connectivity worldwide can be greatly improved using minimum delay topology design. Space optical networks can be improved with respect to network average time delay and with more network stability. Efficient integration of optical satellite constellations with the terrestrial links will have the potential for establishing broadband optical wireless communications and Internet connectivity with *all-optical* technologies for communications *anywhere, anytime*.

6.7 Summary and Conclusions

Design concept technology leading to high-bandwidth, high-data-rate global Internet connectivity and seamless communications using all-optical technology based on integrating laser/optical communication constellation satellites, UAVs, balloons, and air terminals is presented. The analysis is discussed to determine the optimum number of satellites in a given orbit for best communication performance and different communication links for the constellation satellites. Various types of satellites involved in communication constellation: conventional communication satellites, small satellites, microsatellites, CubeSats, and picosatellites [15]: specific orbits: optimum number of satellites for given orbital heights, data rates, and minimizing of latency in the network are discussed. Applications of free-space optical (FSO) communications to design basic architecture of space optical communication networks are also described. A constellation of satellites for laser space communication network using GEO, MEO, and LEO satellites are addressed for both inter-satellite links (ISLs) on the same orbit and interorbital links emphasizing relay satellite communication network using constellation. Finally, constellation designs for LEO mega-constellations for integrated satellite and terrestrial networks to establish global communications and connectivity are also presented.

References

1. A.C. Clarke, "Extra-terrestrial relays", Wireless World, October, 1945, pages 305-308.
2. Arun K. Majumdar, *Optical Wireless Communications for Broadband Global Internet Connectivity: Fundamentals and Potential Applications*, Elsevier, Amsterdam, Netherland 2019.
3. Space: The Ultimate Network Edge, Data Storage Knowledge, October 17, 2016 http://www.datacenterknowledge.com/archives/2016/10/17/space-the-ultimate-network-edge
4. MA Jing, TAN Liying, YU Sinyuan, "Technologies and applications of free-space optical communication and space optical information network", Journal of communications and information networks, 2016.
5. Oltjon Kodheli et al, "Satellite Communications in the New Space Era: A Survey and Future Challenges," IEEE Communications Surveys & Tutorials, Vol. 23, No. 1, First Quarter 2021.
6. E. Leitgeb, S. Sheikh Muhammad, Ch. Chlestil, M. Gebhart, U. Birnbacher, N. Perlot, H. Henniger, D. Giggenbach, J. Horwath, E. Duca, V. Carrozzo, S. Betti, *E-book "Influence of the variability of the propagation channel on mobile, fixed multimedia and optical satellite communications"*, Book Chapter 4 for SJA-2413 "Clear Sky Optics".
7. Vincent W. S. Chan, "Optical Satellite Networks," Journal of Lightwave Technology, Vol. 21, No. 11, November 2003.
8. L. Wood, "Network performance of nongeostationary constellations equipped with inter-satellite links," M.Sc. thesis, University of Surrey, Surrey, U.K. http://www.ee.surrey.ac.uk/Personal/L.Wood
9. B. Chidhambararajan, V. Jawahar Senthilkumar, S. Karthik and Dr. S. K. Srivatsa, "Satellite laser communication networks – A layered approach," Proceedings of the 5th WSEAS International Conference on Telecommunications and Informatics, Istanbul, Turkey, May 27-29, 2006 (pp 467-472).
10. Nikos Karafolas and Stefano Baroni, "Optical satellite networks," January 2001, Journal of Lightwave Technology 18(12):1792-1806.
11. Xiupu Lang, Qi Zhang, Lin Gui, Xuekun Hao, and Haopeng Chen, "A Novel Topology Design Method for Multi-layered Optical Satellite Networks," Q. Yu (Ed.): SINC 2019, CCIS 1169, pp. 87–98, 2020. https://doi.org/10.1007/978-981-15-3442-3_8
12. WIRED.com   Science   06.25.2821   https://www.wired.com/story/tiny-satellites-could-help-warn-of-the-next-big-hurricane/
13. Xin Yang, "Low Earth Orbit (LEO) Mega Constellations – Satellite and Terrestrial Integrated Communication Networks", Ph.D. thesis, University of Surrey, November 2018.
14. Ephraim Pinsky, Danna Linn Barnett, Aharon Oren, "Global Coverage for Fast Response Communication between Constellation of LEO small satellites and Earth stations" Proc. 'The 4S Symposium - Small Satellites Systems and Services', ESA-ESTEC, Noordwijk, The Netherlands, 26–30 May 2008 (ESA SP-660, August 2008)
15. Andreas Fredmer, "Inter-Satellite Link Design for Nanosatellites in New Space," Space Engineering, Master's Level 2020, Luleå University of Technology Department of Computer Science, Electrical and Space Engineering.
16. Marc Sanchez Net, Iñigo del Portillo, Edward Crawley, and Bruce Cameron, "Approximation Methods for Estimating the Availability of Optical Ground Networks," J. OPT. COMMUN. NETW./VOL. 8, NO. 10/OCTOBER 2016.

17. Iñigode lPortillo, Marc Sanchez, Bruce Cameron, Edward Crawley, "Optimal Location of Optical Ground Stations to Serve LEO Spacecraft," Computer Science, 2017 IEEE Aerospace Conference. DOI: https://doi.org/10.1109/AERO.2017.7943631
18. Optical Ground Station, DLR, ESA Contr. No.: 14231/00/NL/WK, September 2001.
19. Ibrahim Shaaban Sanad, "Reduction of Earth Observation System Response time using Relay satellite Constellations," Ph.D. Thesis, The University of British Columbia, Vancouver, April 2020.

Chapter 7
Laser-Based Satellite and Inter-satellite Communication Systems: Advanced Technologies and Performance Analysis

7.1 Introduction

Previous chapters discussed the basics and fundamental technology concepts required to establish space-based laser satellite communications for global communication connectivity using free-space optical/laser links and space/aerial/terrestrial and underwater network architectures [1, 2]. The optimized and the most reliable space optical network design is generally based on the topology of the constellation of satellites to provide connections with the minimum possible number of light-path hops, and the capacity of each light path is determined by the number of connections sharing it. In order to achieve a total *end-to-end communication*, inter-satellite links (ISLs) can be very practical and useful to establish communication link from a satellite in a specific orbit constellation-to-inter-satellite-to-ground/underwater terminals. It is therefore obvious that the space network with the constellation satellites equipped with laser/optical transceivers needs to be of very high throughput so that the space network can be connected seamlessly with the terrestrial backbone network. It will also be necessary to integrate the space, terrestrial, and underwater backbone networks to achieve the ultimate communications *anywhere, anytime* worldwide. Furthermore, *space photonics* can be a key technology for providing a platform technology for all satellite systems. This is very much viable in the present and future laser space communications because of the current availability of the unique performance characteristics and advantages of photonics in terms of bandwidth, mass, power consumption, beam size, immunity to electromagnetic interference (EMI), compactness, and portability. Photonics technology is therefore perfectly suitable for developing laser/optical communications for low Earth orbit (LEO) as well as for other satellite constellations operating at very-low- to very-high-altitude orbits. Space photonics will also be considered for telecom payloads to develop 5G/6G applications to boost the space communications industry downstream. Some of the areas where photonics technology plays essential roles in designing communication terminals between satellites or from satellite to ground

© Springer Nature Switzerland AG 2022
A. K. Majumdar, *Laser Communication with Constellation Satellites, UAVs, HAPs and Balloons*, https://doi.org/10.1007/978-3-031-03972-0_7

include requirements of reduced spacecraft resources such as mass and power. The lasercom systems in 1064 or 1550 nm wavelength windows are matured technologies to establish high-capacity laser space networks from the LEO or geostationary orbit (GEO) satellites. The photonic components which are based on already proven and demonstrated terrestrial photonic technologies include lasers, electro-optical modulators, amplifiers, and low-noise photodetectors and therefore are ready to develop space-based systems and subsystems for high-speed communication terminals equipped with laser/optical transceivers with precise acquisition, pointing, and tracking capabilities. For space-based communication applications, however, the photonic components and their integration into the subsystems have to pass rigorous space-level unit tests.

This chapter introduces the developments of concept technologies needed for establishing global broadband communication and connectivity using satellite constellations at different orbits, where each satellite belonging to a constellation will be equipped with laser/optical transceivers for transferring data communication information between them as well as from/to the constellation of optical ground stations (OGSs). In the next few years, satellite entrepreneurs plan to cover the Earth with thousands of small/miniature satellites to beam low-cost, ubiquitous broadband service, which also includes satellite Internets to remote locations worldwide.

A number of companies have proposed recently to develop and place constellations of communication satellites in low Earth orbit (LEO). The proposed future plans to add a very large number of satellites in LEO include SpaceX to add 11,000 more in the building of its Starlink mega-constellation with permission filed with FCC for another 30,000 satellites and the companies like OneWeb, Amazon, and GW (a Chinese state-owned company) with similar future plan.

7.2   Research Advances in Optical Satellite Networks Relevant to Constellation Designs

One of the essential conditions of achieving continuous, uninterrupted communication and connectivity is to correctly and efficiently design the optical space networks which include GEO, MEO, LEO, and small/nanosatellites at various altitudes and aerial, terrestrial, and underwater terminals creating communication nodes. This is actually a very complex task specially to establish all-optical technologies to accomplish global connectivity anywhere, anytime. However, ongoing research and developments are being done worldwide contributing toward the success of the technology. This section briefly summarizes some of the advances in this area with specific requirements and techniques and methods used to solve them.

7.2.1   *Novel Optical Two-Layered Satellite Network*

Recently, researchers described [4] a novel technique for optical two-layered satellite network of LEO-MEO to satisfy the precise requirement for acquisition, pointing, and tracking (ATP) in optical satellite communication with a zero-phasing factor. Their research shows lower layered satellite constellation to have the stable logical topology of MESH and zero-phasing factor, where the ranges of azimuth and elevation are much lower than those of nonzero-phasing factor. The concept is thus very much applicable for inter-satellite links (ISLs). MEO satellites at two different orbits can therefore be incorporated with LEO to compensate bad coverage for the equatorial region of LEO. The results of computed two-layered satellite spacing connections and coverage performance are discussed with simulation. The concept technology for the two-layered optical satellite network has the potential of 99.9% of global coverage.

7.2.2   *Satellite-Aided Internet Revolution*

Recently reported future technological concept of "metaverse" [5] involves blending the physical world with the digital one that might have a potential application in the worldwide optical wireless communications. This will be particularly of interest to transmit and receive video images from anywhere in the world to any other remote places, thus combining both audio and video between two locations in the world using the "metaverse" technology concept using LEO or other small satellites to accomplish this. This is a more futuristic concept but has the potential for communicating between two remote places.

7.2.3   *Nonmechanical Electro-Optical System (EO) Laser Beam Steering Technology* with No Moving Parts

An entirely nonmechanical electro-optical (EO) laser beam steering technology, with no moving parts for optical da300 g and 5 W respectively. ta connections through air is reported recently [6]. This is an important technology applicable to FSO communication module for small satellite applications characterized by extremely small values of size, weight, and power consumption (SWaP)—approximately 300 cm³, 300 g, and 5 W, respectively—and is capable of 500 mW, 1 mm short-wave infrared (SWIR) beam over a field of view (FOV) of up to 50° × 15°. The range can however be increased by adding polarization gratings. Thus, the device can be easily integrated with the laser communication system on a satellite system. A basic concept of a LEO-distributed *smallsat* can thus be enabled by the nonmechanical laser beam steerers, and the bidirectional communication links can be redirected over a wide range of FOV. The researchers also presented the results of a design effort for this technology for future use over the long distances needed for satellite communication for demonstrating optical power-handling capabilities to provide >W of time-averaged power output as well as to provide unique schemes to close the loop on the pointing angle to provide "self-calibrated" high pointing accuracy.

7.2.4   Cross-Link Laser Beam Problem on CubeSats: Feasibility of Using the MEMS-Based Fast Steering Mirror

Recently, the fine beam pointing problem on CubeSats has been addressed and the feasibility of fine beam pointing has been demonstrated using MEMS-based fast steering mirror (FSM) [7]. Lasercom can improve inter-satellite links (ISLs) for resource-constrained CubeSats with a typical dimension of 3U CubeSat being 34 cm × 10 cm × 10 cm with less than 5 kg mass and about 10 W of available orbit average power. In dealing with CubeSats using lasercom for power-efficient high-rate ISL, some of the design issues, challenges, and requirements are as follows: (a) increasing the optical antenna for ISL is not as feasible a solution because the size and weight are limited for CubeSats (constrained SWaP and cost); (b) the pointing loss of the signal less than 3 dB (to point a beam to a satellite needed to be within its full width at half maximum (FWHM) angle); and (c) the pointing requirement for the CubeSat lasercom inter-satellite cross-link needs to be <12 arcsec for an aperture diameter of 20 mm [7]. The research reported is for each CubeSat to support a full-duplex laser communication link transceiver operating at 1550 nm for inter-satellite communication. The thesis discussed research on the development and analysis of a series of algorithms in order to solve the cross-link laser beam pointing problem on CubeSats and demonstrated the feasibility of using the MEMS-based FSM validating with hardware for fine beam pointing module. A new estimation algorithm using Kalman filtering technique has been reported. An extremely fine pointing system is necessary for the CubeSats because of the requirement of extremely narrow beam requirement for high-rate laser communication link. The technology uses a miniaturized microelectromechanical system (MEMS) and fast steering mirror (FSM) for the feedback control loop to achieve a goal of 20 Mbit/s link at 25–1000 km cross-link range for a high-speed and secure ISL for CubeSats.

7.2.5   *Visible Light Communication (VLC)-Based Inter-satellite Communication System*

Recent paper reported the development of communication subsystem for enhancing inter-satellite communication (ISC) for small satellites by optimizing size, weight, power (SWaP), and cost [8]. Their proposed research is based on inexpensive LED-based visible light communication system satisfying the SWaP and cost constraints suitable for small satellites to establish ISL for achieving high data rate performance. In particular, pico/nanosatellites (category of CubeSats) are considered taking into account background illumination in evaluating link performance. The important feature of these small satellites is low latency between these satellites to provide improved availability for communications application and autonomous operations. When considering downlink communications between LEO satellites and ground stations, these small satellites such as CubeSats can provide reliable mesh of nodes capable of relaying data. Compared to laser-based transceivers equipped in small satellites which require an extremely narrow beam of light to be focused within the FOV, the demands and the requirements are relaxed for LED-based VLC. The system model considered 1U CubeSats in direct LOS with distances between the CubeSats to be fixed; the optical filter at the receiver's front end tuned to the deep Fraunhofer line at 656.2808 nm wavelength with a linewidth of 0.4020 nm, LED peak wavelength of 656.2808 nm, a concentrator radius of 2.0 cm, and PIN photodiode with active physical area of 7.84 cm^2 were assumed in their system study. The research addressed a quantitative assessment of solar background illumination on ISL between small satellites including the performance study for both uncoded and coded IM/DD schemes developing VLC-based communication subsystem for multiple small satellite system. The results from the analytical model show the feasibility of achieving a data rate of 2.0 Mbit/s over a moderate link distance of 0.5 km at a BER of 10^{-6}, which uses a transmitted optical power of 4 W, digital pulse interval modulation (DPIM) scheme, and a receiver bandwidth of 3 MHz. A constellation of properly designed CubeSats (or other types of small satellites) will provide a pathway for accomplishing uninterrupted, high-speed optical wireless communications in remote locations when integrating with a constellation of LEO satellites for developing complete space optical networks.

7.3   Technology Concepts for Inter-spacecraft Omnidirectional Optical Communicator for Future Constellation of Satellites

In order to perform and maintain continuous laser communication between a swarm or a constellation satellite, it is absolutely necessary to design optical transceivers equipped with each satellite which are capable of sending and receiving optical communication signals from all directions, between the satellites simultaneously.

Fig. 7.1 Omnidirectional optical antenna and associated transceivers attached with small satellites for achieving multigigabit data rate optical wireless communication. (*Reprinted with permission from the Author and SPIE, 2018* [9])

For establishing optical communications between the constellation of LEO satellites to the downlink optical ground stations (OGSs), it is often necessary to include a swarm or a constellation of small/nanosatellites for relaying and/or connecting directly to the LEO satellites for example so that communication information from any remote location to any other location can be successfully established. For laser/optical communications, two issues need to be addressed: data rates should be around Gbit/s or more, and optical signals should be transmitting and receiving in all directions to reach the small satellites (e.g., CubeSats) equipped with optical transceivers simultaneously. Inter-satellite omnidirectional optical subsystems have recently been reported [9], which will enable Gbit/s data rates up to 1000 km in free space and will be capable of maintaining multiple simultaneous links with other satellites. Figure 7.1 shows optically interconnected multiple CubeSats to provide potential Gbit/s data. The unique features of these communication subsystems not only enable ISL communications but also provide communication relay links to eventually facilitate developing optical space network nodes for future global Internet connectivity. The constellation satellites in LEO and the swarm of CubeSats can therefore be potentially integrated to establish all-optics technology for global communication connectivity *anywhere, anytime*. The design has been reported [9] to have a small form factor for the CubeSats of four inches, uses a 1 W laser diode transmitter operating at a wavelength of 850 nm, and is capable of 1 Gbit/s data rate at a distance of 200 km distance with a BER of 10^{-9}. The subsystem of optical transceivers contains a set of miniature transmit telescopes consisting of a laser diode, a fixed mirror, and a MEMS. The full sky coverage is provided designing an array of miniature telescopes. Figure 7.2 depicts a 4 in. form factor truncated dodecahedron geometry [9] showing PIN diode detectors and transmit telescopes.

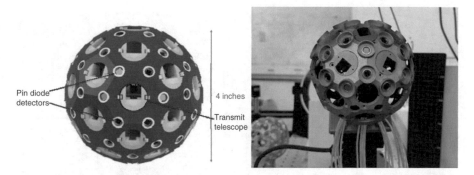

Fig. 7.2 An array of transmit telescopes and pin diode detector to cover the full sky. (*Reprinted with permission from the Author and SPIE, 2018* [9])

The technology developed and discussed in their research paper [9] clearly has the future potential of connecting with hundreds and thousands of small satellites (e.g., CubeSats) to be integrated with thousands of constellation satellites like LEO satellites to establish a global communication and Internet connectivity with all-optical technology.

7.3.1   Advances in Nonmechanical Beam Steering with Polarization Gratings Necessary for Ground/ Satellite-Based Laser Communications

As discussed earlier in this chapter, accurate pointing and tracking capabilities of inter-satellite systems are absolutely necessary for both small satellites such as CubeSats and integrating with a constellation of LEO satellites. However, size, weight, and power (SWaP) requirements are crucial in designing optical transceivers to be installed specially on small satellites such as CubeSats. In addition, cost, reliability, compactness, flexibility, and autonomous capabilities are also desirable for designing a constellation of small satellites to establish communications connectivity worldwide. This subsection discusses one of the important optical components of the subsystems, namely beam steering, necessary for laser satellite communications. Because of short optical wavelength and a finite practical aperture diameter, the pointing accuracy of a laser-transmitting beam needs to be of the order of fraction of milliradian or microradian. Furthermore, because of very high data requirement such multi-Gbit/s, the switching time will also be comparable to close to a fraction of milliradian or microradian. There is a growing demand for rugged, inexpensive, nonmechanical beam steering solutions for ground- and satellite-based laser communications transceivers. Recently, a group of researchers have reported a novel architecture and technology based on polarization gratings for dynamic non-mechanical steering of light over large angles and with large clear apertures [10].

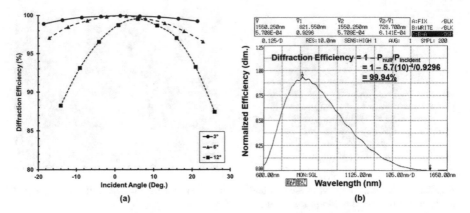

Fig. 7.3 Spectral and angular bandwidth of polarization grating designed for 1550 nm wavelength: (**a**) impact of incident angle on diffraction efficiency with different diffraction angles of ±3°, ±6°, and ±12° and (**b**) spectrum of zeroth-order light transmitted through a 1550 nm polarization grating as a function of wavelength. (*Reprinted and reused with permission from MDPI, 2021* [10])

From the fundamental properties of polarization gratings from a beam steering prospective, it can be shown that polarization gratings offer the unique property of being able to couple 100% of incident light into either +1 or −1 diffraction order when illuminated with circularly polarized light and the retardance is equal to a half wavelength [10]. These polarization gratings with diffraction efficiencies >99.5% are now routinely produced in real manufacturing environments. For lasercom applications operating at 1550 nm, Fig. 7.3 depicts the result of the spectral and angular bandwidth of polarization gratings with different angles and the spectrum of zero-order light transmitted through a polarization grating as a function of wavelength. In designing some laser-based satellite communication systems, it is often necessary to steer a beam waist through a fixed planar aperture in order to aim the beam to another adjacent or nearby satellite in the constellation. With polarization grating steering, steering a beam waist through a fixed planar aperture can also result in a foreshortening of the beam in the axis steering. Figure 7.4 shows the foreshortening of beam waist during steering from a fixed planar aperture where a progressive foreshortening of the beam waist along the direction of steering occurs. This fact therefore needs to be considered when designing the steering system in a small or other laser satellite constellation. Finally, the switching solution is also an important design factor in any polarization grating steering assembly. The technology concept definitely paves the path for accomplishing successful inter-satellite and satellite-to-ground laser communications. Some of the defining characteristics of this beam steering architecture are also discussed in their report. For free-space laser communication, dynamic optical beam steering is a critical optical technology specially for free-space optical links for satellite-to-satellite communications specially when SWaP-constrained applications are needed.

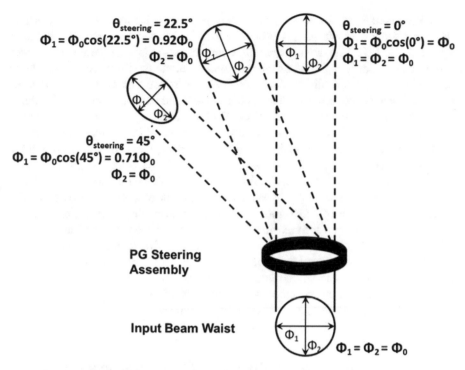

Fig. 7.4 Foreshortening of beam waist during steering from a fixed planar aperture where a progressive foreshortening of the beam waist along the direction of steering occurs. (*Reprinted and reused with permission from MDPI, 2021* [10])

7.3.2   Chip-Based Nonmechanical Beam Steering Technology for Microsatellite Laser Communication Pointing System

A recent paper [11] discusses the feasibility of potential microsatellite laser communication system using nonmechanical beam steering to transmit beacon light. The concept technology was developed and demonstrated by the US Naval Research Laboratory (NRL) and is referred to as SEEOR (Steering Electronic Rejectors) [12] that require no mechanical equipment. This is a chip-based optical technology where the output direction of the input mid-wave infrared (MWIR) laser can be controlled in two dimensions providing greater steering capability and higher scanning speed than the conventional method, which uses expensive, cumbersome, and inefficient mechanical system totally unsuitable for microsatellite applications demanding critical SWaP requirements. The NRL's steerable electro-evanescent optical refractor is based on a slab waveguide design with light mainly confined in a passive high-index core and transiently coupled into a tunable LC upper cladding [11, 12]. A proper design of the electrodes and substrate tapers allows the index change when a voltage applied to the LC changes the refractive index in the waveguide, which can be interpreted as high speed and consecutive steering in two direct ions. Results of SEEOR devices have been shown to provide up to 270° of 1D steering, the angular field of regard (FOR) of up to 50° × 15° for steering in 2D, and high speed of ~60 kHz. The devices are very compact at ~6 cm³, and very small power requirement of only mW makes them extremely attractive for microsatellite laser communications applications. Semiconductor lasers are generally used for laser communications between satellites, which satisfies the requirements of both miniaturization and low power consumption. Although light sources at 800 nm wavelength have been widely used in the past, the current trend for lasercom applications suitable for communication laser transmitters is to design and develop satellite laser communication system at 1550 nm wavelength. The devices and the optical components at 1550 nm are a mature technology, and therefore a chip-based nonmechanical beam steering technology as described above is required to operate at the lasercom wavelength of 1550 nm. Since the nonmechanical steering technology uses short-wave infrared and near-infrared waveguides within the wavelength region of 1000–3000 nm, it is in line with the laser communication requirement of 1550 nm [11, 12]. In conclusion, there is potential for developing nonmechanical chip-based components satisfying the requirements of laser sources, for the range of adjusting beam angles for both the azimuth and pitching direction of rotation range, and for achieving beam pointing accuracy better than 150 μrad.

7.3.3 Laser/Photodetector Arrays for Optical Communications Application in Constellation of Small Satellites (CubeSats)

A fine pointing of laser beams is required for inter-satellite links (ISLs) in small satellites such as CubeSats or between two small satellites at two different low orbits. A method of fine pointing of laser beams has been discussed and a computer simulation to verify the concept technology using novel optical system design [13]. The method has been used by the Aerospace Corporation for CubeSats in LEO at 450 km in NASA's Optical Communications and Sensors Demonstration Program and is developed for application in CubeSats at 100 km at low lunar orbit. The new pointing method discussed by the author combines a novel design of a lens system with an array of vertical-cavity, surface-emitting lasers (VCSELs) and a photodetector array. Some of the features of the unique design concepts include the following: the divergence of the output laser beam can be reduced to 0.014°, the design uses laser array for fine pointing of the laser beams with no moving parts, and reaction times to pointing and changes and vibrations for the proposed electro-optical system are on a nanosecond timescale (much faster than MEMS-based fast steering mirror). For laser communications applications for CubeSats, there is also potential for achieving optical multiple access (OMA) to enable to communicate simultaneously with optical ground stations at different locations as well as to communicate with other satellites in nearby constellation. Using wavelength division multiplexing technique, the data rate transmission can also be increased using this proposed

VCSEL/photodetector arrays. A closed-loop transmission can be accomplished without using optical amplifier employing small-signal current modulation. This electro-optical component is therefore very attractive to be used for laser CubeSats communications, where S laser communications for 10 WaP is also a prime requirement. Note that for applications to LEO at 450 km, or to very low Earth orbit satellites (VLEO) such as CubeSats at around 100 km or less, the atmospheric turbulence effects also need to be considered. The atmospheric turbulence near the ground can be severe to ultimately affect both the LEO and CubeSats to ground station communication for downlink communication. This will particularly be important when the design of the laser satellite communications includes terrestrial links for creating space optical networks for example in designing satellite Internets covering remote locations. Future electro-optical systems for laser satellite communications with all-optical technology will require compact chip-based adaptive optics (AO) modules to be included in the laser/photodetector array design for both LEO and CubeSats communications designs. Figure 7.5 shows a VCSEL/photodetector array with same pitches for the VCSEL clusters and the photodetectors and an actual fabricated VCSEL/photodetector pair [13]. When all laser beams use a 1550 nm wavelength for both incoming and outgoing beams, a possible solution to having the same wavelength might need to have two lens systems, one lens system for received beams and the other for transmitted beams.

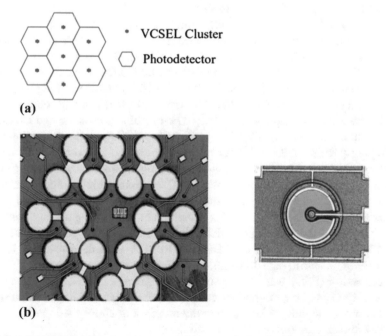

Fig. 7.5 (**a**) A VCSEL/photodetector array with same pitches for the VCSEL clusters and the photodetectors, and (**b**) an actual fabricated VCSEL/photodetector pair. (*Reprinted with permission from the Author and SPIE, 2021* [13])

7.3.4　Advanced Development of Engineering Breadboard Model for LEO and GEO Constellation-Based Optical Communications at 25 Gbit/s

The approvals of the Federal Communications Commission (FCC) for a number of recent Internet constellation programs such as OneWeb, SpaceX, Telesat, and LeoSat have prompted to develop advanced technologies for creating new ways to provide satellite-based broadband access. One of the main reasons for future global Internet traffic will be obviously the tremendous increased numbers of customers, both individuals and machines, across the world. Satellite communication is the only way where various constellations at different orbits can connect users anywhere worldwide including remote locations. This book has addressed the fact that free-space optical communication (FSOC) offers the only solution for establishing two-way (duplex) communications between space-based platforms and optical ground stations (OGSs). Techniques for establishing optical wireless communications for broadband global Internet connectivity have been discussed in detail [3]. This subsection discusses the research and development of a communication system unit and reports the advanced optical transceiver and the modulators implementing the on-off keying (OOK) and differential phase-shift keying (DPSK) formats for providing very-high-speed (~25 Gbit/s) data transmission [14]. The research is developed by Airbus Defense and Space in France and is very suitable for incorporating in LEO, GEO, and small satellites like CubeSats where reduced overall size, weight, and power (SWaP) are important considerations. The research will also be significant for establishing inter-satellite links (ISLs). One of the key technologies for the optical feeder links (FOLC project Feeder Optical Link for Constellation [14]) is also discussed. The engineering breadboard is composed of digital processing u-optical booster: the nit, the optical transceiver (Tx/Rx), and the optical booster (a high-power optical amplifier (HPOA)). Their results discuss the two modulation schemes, nonreturn-to-zero (NRZ) on-off keying (OOK) and differential phase-shift keying (DPSK) at 25 Gbit/s for potential laser satellite communications. The system reported consists of a digital processing unit, an optical transceiver (Tx/Rx), and a 5 W optical booster. Some nonlinear effects have started to appear inside the 5 W booster. Wavelength division multiplexing (WDM) technology was used to increase the data transmission for 3×25 Gbit/s channels. The average sensitivity reported for NRZ-OOK and NRZ-DPSK links at 25 Gbit/s, respectively, are -36.8 and -40.2 dBm for a bit error rate (BER) at 10^{-3} [14]. The demonstration technology can be applicable to both downlink and uplink optical communications. Forward error correction (FEC) was also assumed to improve the BER up to quasi-error operation. FEC can be an important consideration when designing such an optical unit to improve the link performance in the presence of atmospheric turbulence for both downlink and uplink optical transmissions.

7.3.5 Advanced Technology of Compact, Optical Components for Small Satellites for Optimized Power Transmission and Reception Future Efficient Optical Data Center: Ultrafast Optical Circuit Switching for Photonic Integrating Platform

A recent report [15] demonstrated ultrafast optical circuit switching using a chip-based soliton comb laser and a completely passive diffraction grating device. The architecture performing wavelength switching is particularly important to laser satellite communications, where different servers can be connected with different wavelengths of light with a great advantage of limiting the need for electrical switches and optical transceivers placed in space/aerial platforms including small satellites. The switching element, essentially a glass prism, can switch and route signals to destination servers by dispersion. The new emerging technology does not require electrical-to-optical conversion, which the current network architectures rely on using electrical chips. The chip-scale indium phosphide-based optical amplifiers perform the switching between different wavelengths. This all-optical technology will be extremely useful for minimizing the cost and power overhead and for advancing the applications requiring growing data rates and AI for 6G and more. The proposed architecture optical microcombs provide the multiwavelength source for communication coherent carriers, whereas optical amplifiers and arrayed semiconductor waveguide gratings perform switching separating or combining different wavelengths of light [15]. For laser-based small satellites, these optical technologies serve the purpose of high-data-rate switching, lower cost and power, compactness, and flexibility required for both inter-satellite and terrestrial communication links in satellite applications.

7.4 Some Advanced FSO Technologies and Components Relevant to Laser Satellite Communications

This section discusses a few more recent advanced optical devices, components, and subsystems, which have potential uses for designing and developing laser-based satellite communications technologies for very-high-data-rate transfers *anywhere, anytime* to space/aerial, terrestrial, and underwater terminals. Free-space optical technologies are definitely creating future communications infrastructure for global communication systems using cutting-edge technologies. Research, development, and demonstrations are being continuously accomplished by various laboratories, universities, and industries toward these goals. Cutting-edge free-space optical communication (FSOC) terminals are essential to ISLs between small satellites as well as can serve as optical feeder links for large, very-high-throughput FSO communication links for GEO communication satellites.

7.4.1 High-Power (100 W) Optical Amplifier for WDM in Satellite Communication

A 100 W optical amplifier for a WDM optical communication system has been discussed in a recent conference [16]. The amplifier can handle 10-channel laser communication system with enhanced wall-plug efficiency in the 1 μm wavelength range. The technology is based on all-fiber technology with a desired efficiency of 30% or more. The high-power optical amplifier will be essential for optical signal transmission for long distance and is applicable for inter-satellite laser communications for small, LEO, and GEO satellites to provide high signal-to-noise ratio and hence to achieve high data rates in multiple Gbit/s or much more as needed.

7.4.2 Single Optical Amplifier for Both 1064 and 1550 nm Wavelength for Laser Satellite Communications

A single optical amplifier applicable to both optical communication wavelengths of 1550 nm (lasercom) and 1064 nm (potential greater efficiency for small SWaP requirements) is discussed [17]. The authors report the development of the optical amplifier using an ErYb-only fiber amplifier and a hybrid Yb/ErYb fiber amplifier. The dual-band optical amplifier can be useful in optical communication systems for inter-satellite and satellite-to-ground communications, where one wavelength can be used for target detection and tracking and the other wavelength for communication carrier.

7.4.3 Optical Circulator for Free-Space Optical Communication

Free-space optical circulator, a useful device for practical purpose for space communications, is proposed and tested to transmit and receive paths in the optical C band using magnetless Faraday rotators (MFR) [18]. The C band (conventional band) normally ranges from 1530 to 1565 nm. The searchers claim that the system combines/splits the optical paths close to 100%, and the wavelengths can be separated by a few 10s of nm. When integrated in the satellite-based optical terminals, the device can be easily operated with arbitrary wavelengths and polarizations and can be designed for efficient satellite laser communications.

7.4.4   Multiplane Light Conversion Module for Atmospheric Turbulence Mitigation for Satellite-to-Ground Laser Communication

A novel atmospheric turbulence mitigation technique for satellite-to-ground FSOC as been recently demonstrated without using conventional adaptive optics (AO) [19]. The module uses an 8-mode multiplane light conversion (MPLC) connected with a photonic integrated chip (PIC) to collect an atmospheric turbulence-induced distorted beam and converts it into a fundamental mode propagating in a standard single-mode fiber. The concept has been tested by the researchers for various aperture diameters and atmospheric distortion parameters to simulate from turbulent-phase plates. For designing efficient high-throughput communications suitable for satellite-to-ground FSOC systems, this turbulence mitigation technique has potential applications in future.

7.4.5   Space Optical Switch Development: New Concepts of Photonic-Based Switch for High Data Throughput up to 1 Tbit/s

Researchers have recently developed space optical switch capable of high data throughput up to 1 Tbit/s for GEO and for low-cost LEO constellation satellites [20]. The optical switching is one of the most important key photonic technologies to provide higher capacity and high payload flexibility in optical communication with added advantages of redundancy functionalities and optical interconnecting. Furthermore, recent developments of LEO satellite constellation will be extremely helpful to use very-high-data-rate-capacity optical switches for enhancing laser links for full optical signal-based telecom satellite, which require reduced price/size/mass reduction constraints. Some of the obvious advantages using optical switch for photonic payloads include flexible allocation of gateways, redundancy functionalities, and large-scale optical interconnection [20]. The report discusses the innovative principle used in mirror reflection and basically has the following components: mirror, 2D piezo actuator, fibers, and digital position control loop. Each input fiber with optical collimator generates a parallel beam where the 2D piezoelectric actuator controls the collimator orientation using integrated position sensor. After the output collimator focuses the beam on the fiber end, input collimator and output collimator are aligned. The wavelength of operation used in the report for space application was 1.55 μm. The laser communication satellite applications and concept technology have the scalability up to 384 × 384 with modular design for future laser communication satellites including LEO and small-satellite constellation-equipped optical switch matrix. *The all-optical switches are one of the most important space optical communication modules for future architecture for photonic payloads for space laser satellites and constellations.*

7.4.6   Development of Laser Communication Terminals for Satellite and Terrestrial Networks

Laser communication terminals for 10 Gbit/s space communications with a 1.5 μm band laser beam between satellites and optical ground stations (OGSs) for establishing space/aerial and terrestrial optical networks have been recently reported [21] by the researchers at the National Institute of Information and Communications Technology (NICT) in Japan. The technology demonstrations of laser communication networks with satellites/airborne and terrestrial including underwater FSOC terminals will lead toward establishing future high-speed and high-secured laser communications to remote locations.

7.4.7   Modulating Retroreflector (MRR) Technology to Provide High-Capacity Fronthaul/Backhaul Optical Links for UAVs/Drones and Base Stations (BSs)

Original research on (NRL) multiple quantum well (MQW) modulating retroreflector (MRR) started many years ago at the Naval Research Laboratory (NRL). The MRR concept technology involves coupling an optical corner cube retroreflector and an electro-optic shutter to provide two-way optical communications using a laser source and a pointer-tracker on only one platform. The detailed description and operating principles as applied to UAV free-space laser communications for UAV-to-ground base as well as between UAVs are already discussed [22–24]. Modulating retroreflector FSO communication terminal on UAV and a new methodology of using multiple UAVs in a cooperative swarm mode and adaptive beam divergence for inter-UAV FSO communication, networking architectures, reliability, and appropriate modulation schemes are also discussed. Tracking in moving platforms and other challenges are addressed in their papers as follows: variation in receiver beam profile of the FSP link and variation in received optical power due to constantly changing transmitter/receiver separation. Incorporating UAVs (or small to nanosatellites) with base terminals and free-space optical communication (FSOC) has potential applications for the upcoming sixth-generation (6G) cellular networks for IoT or machine-to-machine (M2M) uses. UAVs or small satellites can be equipped with MRR subsystems as mentioned earlier and can achieve automatic acquisition, tracking, and pointing (ATP) and overcome the constraints of size, weight, and power (SWaP). However, atmospheric turbulence can still put a data rate limit in transferring and receiving optical beams between UAV and a base terminal specially for a 100 m altitude from the ground. The effects can be severe for the first few meters from the ground at the atmospheric boundary layer where the strength of turbulence parameter, C_n^2, is high and can be randomly unstable. This can affect transmitting and receiving the laser communication signals to and from the UAVs. The atmospheric parameters of interest may include angle-of-arrival

(AoA) fluctuations at the receiver, turbulence-induced channel coherence time and fading duration, and received signal-to-noise ratio (SNR) to determine the achievable bit error rate (BER). For MRR application, it is also important to consider the turbulence characteristics of the direct and then the modulated retroreflected beams along the same path. Channel statistics of drone-to-ground retroreflected FSO links have been discussed in a recent paper [25], which reports experimental results of interest to design a MRR-based FSO system for hovering altitudes of 15 and 20 m, over roundtrip distances of 201 and 204 m. Laser optical wavelength of their report for designing fine tracking system was 976 nm with the divergence of 0.04 rad. Some of their results include AoA at the optical ground station's telescope entrance, probability density functions (PDF) of the beam centroid displacement, DF of the normalized received power, and probability of fade. The paper clearly provides important design parameters useful in developing MRR-based UAV/small satellite-to-ground optical communication system in the presence of atmospheric turbulence. This will pave a pathway for fronthaul/backhaul links in the future 6G cellular networks.

7.5   Satellite-Based Quantum Communications

Free-Space Quantum Communications
Quantum communication is a very fast field growing in importance whose applications are expected to play a vital role in commercial, domestic, and defense applications. Free-space quantum communications infrastructure is therefore important to consider recent satellite-based communications technologies requiring various scenarios such as ground-to-satellites at various orbits, inter-satellite links (ISLs), and other aerial/terrestrial terminals for transferring and receiving very-high-speed information data in the most possible *secure* way existing today. In short, "speed," "portability," and "(almost) unbreakable security" are the key parameters need to be considered in order to design and implement effective free-space quantum communication systems. Recent successful experiments for transmitting and receiving "photons" over long distances paved the way for potential feasibility of satellite quantum communications [42-44, 46] from ground to GEO/LEO/small satellite or between satellites (inter-satellite configurations).

Brief Summary of Basics of Satellite-Based Quantum Communications
Some basics of free-space and atmospheric quantum communications are already discussed [23, 26], and the interested readers are referred to those research papers and book(s) for detailed discussions on the subject. It is obvious that secure communications for sensitive information over the Internet is very critical. Long-distance free-space atmospheric quantum communications will therefore play a critical role in extending the quantum Internet to global use. Satellites (GEO and LEO, and small/micro), HAPs, UAVs, and aerial and free-space underwater terminals will have to be included for quantum information to be teleported through

mobile information teleportation networks. Basics of quantum physics developments to add a physical layer of quantum security beyond classical communications capabilities to free-space and atmospheric communications are briefly summarized for background material. This section also addresses some of the available and demonstrated communication technologies required to implement in a real satellite-to-ground system to achieve secure quantum communications.

Some of the elements of quantum communication systems relevant to long-distance space (satellite/aerial) to ground are defined below (the readers are encouraged to review the excellent papers in the references [23, 26–28] for more detailed discussions on quantum communication-based secure communication and demonstration experiments):

- *Quantum channel*: A quantum channel capable of transmitting both quantum and classical information. For example, a quantum information is the state of a *qubit* (the basic unit, the quantum version of the classic binary bit), whereas an example of classical information is a text document transmitted over the Internet [27]. The quantum channel can therefore be considered as a pipeline to carry quantum information. Free-space transmission of a single photon can implement a quantum channel experimentally.
- *Quantum teleportation*: This can be defined as a process where a sender wishes to transmit an arbitrary quantum state of a particle to a possibly distant receiver: this requires a quantum channel for the transmission of one particle of an entangled state to the receiver [27].
- *Quantum networks for communication*: Quantum networks make it easier to transmit *qubits* information between physically separated quantum processors, which can perform quantum logic gates on a certain number of *qubits*. Quantum Internet can intraconnect local quantum networks supporting many applications. Information can be transmitted between the remote processors by creating *quantum-entangled qubits* [28]. Furthermore, in quantum Internet protocols such as quantum key distribution (QKD) needed in quantum cryptography, only a single qubit at a time is prepared and measured by the processors [28]. For quantum cryptography purposes, only one qubit is required because quantum entanglement can be realized between just two qubits. Optical networks and photon-based qubits are required to design long-distance quantum networks through the atmosphere for satellite to ground for transmitting quantum information. A quantum satellite capable of long-distance entanglement distribution has been recently demonstrated.

In a classical channel for a free-space optical network, sensitive data is typically encrypted and then sent across atmospheric channels together with the digital "*keys*" needed to decode the information. Both the data and the keys are sent as classical bits, for example via optical pulses representing *1s* and *0s*. The process however is vulnerable to security, and hackers can break the code to read the sensitive data. Quantum communication on the other hand allows particles (for example, photons of light), which are called *qubits* in quantum communications to transmit data along the communication path allowing to take on superpositions of *1* and *0* simultaneously. If a hacker attempts to observe the qubits in transit, immediately the quantum states of the qubits "collapse" to either *1* or *0*, thereby leaving the sign of tempering activity. Quantum key distribution (QKD) is a very efficient process to design ultra-secure networks for transmitting highly secure sensitive information. There are many protocols to implement QKD including BB84, which is widely used. This section on long-distance atmosphere emphasizes developing quantum secure communications (QSC) relevant to satellite/aerial long-distance communication in the presence of atmospheric turbulence and scattering causing decoherence and attenuation of laser communication signal. As a simple example, as depicted in Fig. 7.6, let Alice (sender) want to send Bob (receiver) data information securely through the atmospheric channel. Alice first creates an encryption key. Polarization states of the qubits of the encrypted key represent the individual bit values of the key. The authenticity of the encryption keys, i.e., both Alice and Bob have the same key, is determined by a process called "key shifting" that compares the measurements of the state of a fraction of some of these qubits collapsing because of decoherence due to atmospheric propagation [29]. If the error rate is not acceptable, Alice and Bob conclude that the key is intercepted by a hacker and generate a new secure key to decode the information. Alice then encrypts data with her established secure key in classical bits to Bob who decodes the information with the secure key. A recent demonstration of an integrated space-to-ground quantum communication network that combines a large-scale fiber QKD link and two high-speed satellite-to-ground free-space QKD links has been reported [30]. The integrated link was extended to a remote node more than 2600 km away so that any user in the network can communicate with any other up to a total distance of 4600 km.

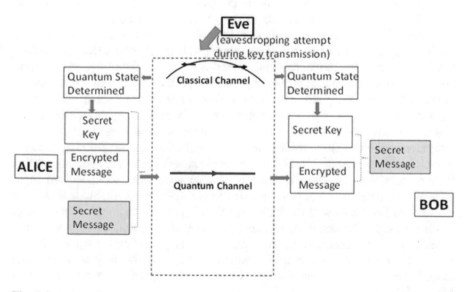

Fig. 7.6 A simplified diagram showing free-space quantum communication using quantum key distribution (QKD)

7.5.1   Satellite-Based Satellite Quantum Communications: Challenges and Progresses for Implementing QKD over Long Distance Across Free-Space Channels

In order to access the QKD security for satellite-based laser communications, the basic parameter for open-air quantum key distribution is the effect of atmospheric attenuation on the polarized encoded photons that needs to be evaluated. A few years ago, researchers presented [32] a full implementation of a QKD system with a single photon source, operating at night. The free-space propagation path of 30 m was tested, and the experimental setup was operated at night and corresponds to the implementation of the BB84 protocol. The free-space quantum channel between Alice and Bob and the Internet as the classical channel were used. They compared and demonstrated the number of shared secured bits/pulse for a weak coherent pulse source versus a true single photon source. For example, to achieve the same number of shared secure bits/pulse of 10^{-5}, only 10.7 dB attenuation on quantum channel for a weak coherent pulse source was needed, whereas a larger attenuation of 12 dB could be tolerated on quantum channel for a true single photon source. The results thus show the advantages of single photon source over the equivalent weak coherent pulse source demonstrating that QKD with single photon source outperforms QKD with weak coherent pulses when transmission losses exceed 10 dB [32]. Note that their experiment was conducted using a homebuilt pulsed laser at 532 nm with a 0.8 ns pulse duration to excite single nitrogen vacancy color center (the single photon source was based on temporal control of the fluorescence of single-color nitrogen vacancy center in diamond nanocrystal). However, the concept of using single photon source at the telecom wavelength range around 1550 nm for free-space QKD from satellites is promising to continue in the near future [32].

Recently, researchers demonstrated long-distance and secure measurement device-independent quantum key distribution (MDI-QKD) over a free-space channel [30]. In order to achieve the goal to establish the global scale quantum secure communication network, it is important to extend the distance of implementing QKD under any atmospheric condition such as turbulence and signal attenuation. An efficient MDI-QKD protocol can be useful in addressing this challenge on detection at once and can improve the performance and security of QKD implementations on real devices by including decoy states. Satellite-based QKD implementations between satellite and ground can therefore increase the secure communication distance considering free-space channels with lower transmission losses and negligible decoherence in space. Atmospheric turbulence-induced amplitude and phase fluctuations limit how well the indistinguishability can be maintained in terms of spatial, timing, and spectral modes between independent photons because the atmospheric turbulence generally destroys the spatial mode between independent photons [30]. Their results are useful which could pave the way to developing long-distance secure communication by properly implementing satellite-based MDI-QKD.

7.5.1.1   Quantum Key Distribution (QKD)

LEO-to-Ground Link in the Presence of Atmospheric Turbulence
and Scattering

Worldwide free-space optical (FSO) quantum communication (QC) network requires integration of satellite-based optical links and fiber-based networks. Quantum signals suffer signal losses due to atmospheric turbulence and attenuation by scattering during the propagation along an atmospheric FSO link. For successful satellite-based QKD missions, it is therefore necessary to evaluate the atmospheric effects on designing and implementing the QKD protocols. One of the parameters to quantify the effects is to calculate the achievable key rates using different implementations of QKD. The nonuniform free-space link between the satellite and the ground station consists of (1) the atmospheric turbulence along the slant path (including the boundary value turbulence layer close to the few hundred meters near the ground) between the ground station and the satellite link within the thickness of the atmosphere (2) and the slant path between the satellite and the top of the atmosphere. Note that these slant-path distances between sender and receiver are different from the height of the satellite from the ground station and the thickness of the atmosphere. Recent research [31] addresses a scenario of satellite-based link with LEO satellites above 500 km for evaluating the key distribution in the interval of 500–2000 km altitudes of the satellite, the effective thickness of the atmosphere of 20 km, and the corresponding zenith angles in the interval of $[0°, 80°]$. The authors considered the elliptic beam approximation to calculate the transmittance as a function of the elliptic beam parameters such as the beam centroid coordinates, the principal semiaxis of the received elliptic beam profile, and the radius of the receiving aperture. For the satellite link model of the atmosphere, both turbulence and scattering effects along the propagation path (appropriate for both uplink and downlink) were taken into account for determining the transmittance. Both downlinks and uplinks for propagation are considered in order to evaluate the rates of QKD. In the quantitative determinations of the QKD, the uplink and downlink probabilities of distributions of transmittance (PDT) of the fluctuating optical link were taken into account. There are three parameters which are relevant to establish the accurate feasibility of satellite-based QKD mission: (1) the exact protocols to be selected which apply to this long-distance satellite to ground scenario to handle the atmospheric turbulence effects, and the background noise; (2) quantify the expected secret key rates of a QKD protocol chosen for the scenario; and (3) the corresponding quantum bit error rate (QBER) averaged over the probability distribution of the atmospheric transmittance. The researchers discussed the performance of QKD implementation for satellite-based FSO quantum communication and computed the expected secret key rates of QKD protocol. The performance of the BB84 protocol was analyzed with polarization encoding, implemented using either a true single photon (SP) source or a weak coherence pulse (WCP) [31].

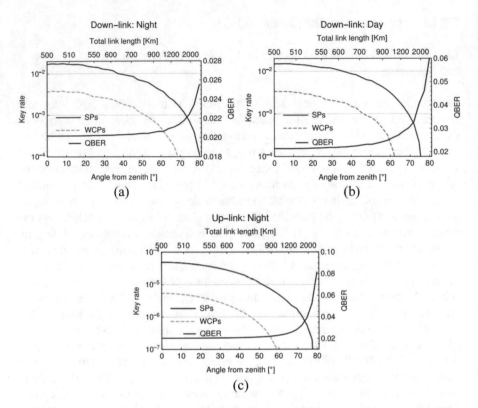

Fig. 7.7 LEO satellite quantum secret key rates generated by the BB84 protocol for various parameters: (**a**) downlink (night), (**b**) downlink (day), and (**c**) uplink (night). (*Reprinted with permission from the Open Access the New Journal of Physics 2019* [31])

The results reported in their recent paper [31] are summarized as follows. The averaged key rate, $\bar{R}$, is given by

$$\bar{R} = \int_0^1 R(\eta)\mathcal{P}(\eta)\mathrm{d}\eta = \sum_{i=1}^{N_{\mathrm{bins}}} R(\eta_i)\mathcal{P}(\eta_i) \qquad (7.1)$$

where

$R(\eta)$ is the key rate at the specific value of the transmittance.
$\mathcal{P}(\eta)$ is the probability distribution of this transmittance.

The integral average is approximated by dividing the range [0, 1] in N_{bins}.

Figure 7.7 depicts the secret key rate for downlinks (night and day) and an uplink (night) for a QKD BB84 protocol with polarization encoding for a true single photon (SP) source or weak coherent pulses (WCPs) for ground-LEO satellite link ranges and for different zenith angles. Figure 7.7a, b shows the key rate for downlinks (night and day) and Fig. 7.7c shows the key rate for uplink (night) together with the parameter, quantum bit error rate (QBER). The QBER is defined as follows [31], assuming that the noisy photons are completely unpolarized:

$$\text{QBER} = Q_0 + (1/2)\left[N_{\text{noise}} / (N_{\text{noise}} + N_{\text{signal}}) \right] \quad (7.2)$$

where Q_0 represents the error rate associated with depolarization in the encoding and the value was assumed to be equal to 2%, and N_{noise} and N_{signal} are the number of photons per time window of noise and signal, respectively. Note that for LEO satellite of total link length of ~640 km, the downlink single photon key rate at night is about 10^{-2} for single photon implementation, whereas the rate is about 8×10^{-3} for daytime link, and the key rate is approximately 4×10^{-5} for nighttime uplink.

CubeSat Quantum Secret Key Rates

The researchers also discussed the secret key rates applicable to nanosatellites (CubeSats). CubeSats are potentially practical to implement and develop worldwide quantum communication network. In this case, normally smaller optics of diameter less than 10 cm are used resulting in smaller transmittance for the uplink configuration [31]. Figure 7.8 shows the key rates achievable for CubeSats in the downlink and uplink configurations.

Inter-satellite Free-Space Optical Quantum Key Distribution (QKD)

Global optical communication systems require links between satellites and space stations or spacecrafts. Both inter-satellite links (ISLs) and interorbital links (IOLs) will be needed to establish global broadband Internet connectivity to remote locations. This section discusses briefly inter-satellite free-space optical quantum key distribution (QKD) necessary for secure quantum communications.

Some other related research efforts and development in quantum key distribution (QKD) for quantum networks.

7.5.1.2   Silicon-Photonic Chip to Integrate Detectors for QKD Devices

It is evident from the previous discussion that the single photon source and the single photon detector are key elements for quantum key distribution relevant to quantum secure laser satellite communication. The single photon detector is therefore highly desirable for photonic chip integration to design and develop practical and scalable quantum networks. Recently, optimal time-bin-encoded measurements are realized using a superconducting silicon chip used as an untrusted relay server for secure quantum communication [33]. The authors reported that the optimal

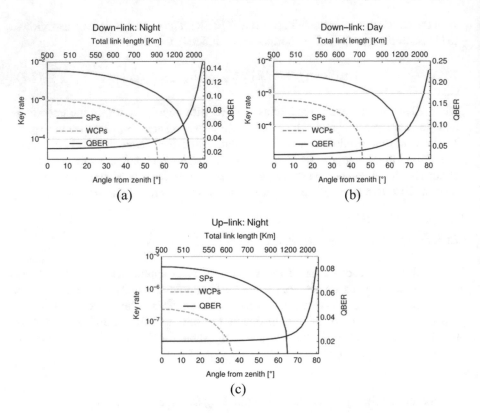

Fig. 7.8 CubeSats' key rate using the BB84 protocol for the same parameters used in the earlier figure as LEO-ground satellite: (**a**) downlink (night), (**b**) downlink (day), and (**c**) uplink (night). (*Reprinted with permission from the Open Access the New Journal of Physics 2019* [31])

Bell-state measurement of time-bin-encoded qubits generated from two independent lasers increased the key rate of measurement device-independent QKD immune to all attacks against the detection system [34]. A QKD network with untrusted relays can therefore be developed. The researchers reported a secure key rate of 6.166 kbit/s over 24.0 dB loss. When integrated with QKD transmitters, the basic concept from their result provides a pathway for future miniaturized, scalable, chip-based, and cost-effective QKD network applicable to a constellation of laser satellites for achieving secure quantum communications between the satellites as well as from satellites to ground stations.

Some of the recent advances in developing chip-based quantum key distribution are reported with recent advances in implementing protocols, photon sources, and photon detectors. Integrated photonics can offer the compactness and the full compatibility with classical devices, which would allow QKD devices to be embedded into classical devices, and has a number of advantages including miniaturization, low energy consumption, and a potential low cost [36]. Future possibility of high-speed modulation of quantum states in standard silicon photonics will be very useful in providing a seamless integration with classical communication channels and transceivers by integrating with current integrated photonic telecommunication hardware. The development in this area will thus provide practical hybrid classical and quantum communication devices in the near future [36].

7.5.1.3   Atmospheric Free-Space Link Strategies for Achieving Potential High-Data-Rate Satellite-Based QKD

Optical satellite links needed to overcome the distance limitations of ground-based laser communications and entanglement-based QKD protocols are required to remove the need to trust the source on the satellite to establish duplex (uplink and downlink) secure communications. Distributed entangled photons from an ultrabright source over a 143-km-long atmospheric free-space link with >0 dB total channel loss are reported [35]. The scenario is comparable to the average downlink loss for a low Earth orbit (LEO) satellite link. The researchers implemented the BBM92 protocol with error correction and privacy amplification and demonstrated secure key rate up to 300 bits per second (bps) ranking among the highest key rates over free-space channel with >40 dB total channel loss [35]. The propagation link between the satellite and the ground station is dynamically (randomly) varying, and the attenuation in a satellite downlink depends on the orbit of the satellite. The transmitter and the receiver can therefore be optimized based on the current link conditions. They obtained a secure key rate of 154 bps over 85 s with an average total channel loss of 46.9 dB and a secure key rate of 300 bps over 15 s with an average total channel loss of 43.5 dB [35]. By adjusting the photon pair rate on the satellite, the number of secure bits exchanged can also be adjusted within a short LEO satellite pass. By measuring the power of the beacon or synchronization laser at the satellite, the time-varying attenuation can be monitored. Future satellite-based QKD space networks will obviously depend on the temporal (and spatial) variations of the atmospheric models between the satellite and the ground used. All these parameters of transmitter and receiver that are discussed and their dynamic adjustments will provide the potential long-distance satellite-ground-based QKD generation for secure space communication.

7.5.1.4   LEO and GEO Quantum Satellite Constellation for QKD Network Based on Trusted Repeaters

Recently, researchers studied the resource allocation problem of a QKD network that combines both GEO and LEO satellite constellations in order to exchange secret keys in a global coverage [37]. QKD can be an optical technology which can be used to automate the delivery of encryption keys between any two points sharing a free-space optical link. However, the power loss of the physical communication channel (propagation loss) increases exponentially with the link distance, which limits the rate of exchanging the keys over a long coverage range using GEO or

LEO constellation satellites. To solve this problem, trusted relays placed on satellite payloads can be used to make the operation of eavesdrop difficult. The researchers in their work considered the BB84 protocol with decoy state in the quantum channels and the relay of secret keys between ground stations performed using trusted relays in the satellites. The satellite QKD network in their model considered the ground station (GS) and quantum satellites as *nodes*, the quantum channels as *edges*, and the amount of available secure keys as *weighs* [37] with the constraints in the GEO-to-GS, LEO-to-GS, and LEO-to-LEO optical quantum links.

- The ground station located globally are equipped with free-space optical (FSO) transceivers so that both the GEO and about two LEO (depending on how close the LEO satellites with each other are, which again depends on how many LEO satellites are deployed in a given orbit) are observed within the field of view of the OGS. Figure 7.9 shows a sketch of establishing global quantum key distribution (QKD) optical space network incorporating GEO and LEO constellation and a large number of optical ground stations (OGSs). A number of LEO satel-

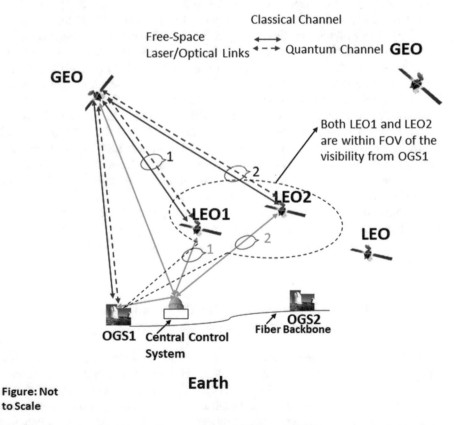

Fig. 7.9 Concept architecture for free-space optical (FS)-based QKD between optical ground stations (OGSs) located in various locations globally using GEO and LEO trusted repeaters. (Figure: Not to scale)

lites' trusted repeaters can be used as long as the LEO satellites are within the view from a given OGS. The figure depicts two types of relays: (1) trusted relays using the GEO and LEO satellites and (2) relay of secret keys between ground stations with the aid of trusted relays of (1). Global QKD can be enhanced by a number of LEO-based inter-satellite relays.

Their technique and implementation can be described in the following way: first of all, a central control system is required with a knowledge of the rate at which the secret keys can be generated on both the ground-to-space and inter-satellite quantum channels so that the most convenient routing on the classical channels can be optimized for the flow of secret keys. Note that all the links of GS-to-LEO, LEO-to-GEO, and GS-to-GEO have both classical and quantum channels. The process thus informs these decisions to the QKD nodes using the control channels [37]. The centralized routing algorithm addressed by the researchers therefore selects the most convenient trusted relays to forward the secret keys between pairs of ground stations. Their results demonstrate that long-distance quantum communication can be achieved using GEO and LEO satellites. There is also potential for developing the repeater-based quantum networks using nanosatellites. The optimization of the topology and the efficient configurations still need to be explored, specially when connecting GEO, LEO, and nanosatellite (CubeSats) global quantum network. The future research in these areas will lead to developing true *quantum Internet.*

7.5.1.5   Modulating Retroreflector-Based QKD in Turbulent Channel

A group of researchers recently discussed a plug-and-play measurement device-independent quantum key distribution (MDI-QKD) with modulating retroreflector (MRR) scheme. The MRR device can compensate the polarization drift during the transmission to provide better performance when implementing the MDI-QKD on a free-space optical channel. The MRR device can help in the self-alignment process on both communication terminals located in airborne/satellite-based and ground terminals. The paper addressed the security analysis for MRR-based MDI-QKD scheme under some classical attacks. Their simulation results indicate the future potential of constructing MRR-based MDI-QKD architectures suitable for moving platforms in creating dynamic quantum network with untrusted relays [38].

7.5.1.6   Air-Sea Secure Quantum Communication: Quantum Key
Distribution for Satellite-to-Underwater Laser Communication

A practical global quantum communication network requires to include an air-sea free-space channel to link the satellite-based quantum resource and underwater terminals. An experimental demonstration of air-water decoy-state quantum key distribution keeping a low-quantum bit error rate less than 2.5% for different distances has been reported recently [39]. Their experimental setup uses four blue laser diodes operating at 450 nm wavelength at the Alice end and a green laser diode at 520 nm for synchronization. A 2-m-long free-space channel was guided to the underwater channel by two mirrors, and the collimated laser beam at the Bob end used BB84 decoding module by mirrors and lens. Local area network was used by both Alice and Bob for classical channel communication. Their results show the quantum key generation rate (pulse^{-1}) of 10^{-5} for an equivalent 345-m-long clean seawater of Jerlov type I. The results provide a key step forward to future practical satellite-sea secure communication. Other similar research works are discussed [41, 45] and verified the feasibility of quantum key distribution (QKD) in a 10 m water channel, where the authors presented an experimental investigation of QKD and decoy-state QKD based on BB84 protocol. The security key rate of 563.41 kbit/s and the quantum bit error rate (QBER) of 0.36% were achieved [40]. When two decoy states were used in the 10 m underwater decoy-state QKD experiment, with an average QBER signal of 0.95%, the security rates reached 711.29 kbit/s. Using a lower count single photon detector (SPD), the researchers predicted the maximum secure transmission distance of 19.2 m which could be increased to 237.1 m in Jerlov type I seawater [40].

7.6   Summary and Conclusions

This chapter discusses current advanced technologies with laser-based satellite and inter-satellite communication systems required to establish global communication systems. This chapter presents the concepts of optical satellite networks relevant to constellation design and covers the establishment of satellite-aided Internet. Some of the device technologies include laser beam steering technology with no moving parts and the MEMS-based fast steering mirror specifically useful for CubeSat constellations. Inter-satellite communication system is addressed for future development of constellation of satellites. Microsatellite laser communication systems requiring pointing system using chip-based nonmechanical beam steering system, and other optical components like laser/photodetector arrays, are also discussed in this chapter. Optical and photonic components such as high-power optical amplifier needed for the development of laser communication terminals for satellite and terrestrial networks are explained. Another technology based on modulating retroreflector (MRR) to provide high-capacity fronthaul/backhaul optical links for UAVs and base stations is also described. Multiplane light conversion module for atmospheric turbulence mitigation for satellite to ground is also presented. Finally, satellite-based global quantum communications and integrated networks are also discussed. Challenges and progresses for implementing quantum key distribution (QKD) over long distance across free-space channels are also specifically addressed in this chapter. Recent developments of implementing QKD for LEO-to-ground link as well as for inter-satellite links in the presence of atmospheric turbulence are discussed and explained.

References

1. Arun K. Majumdar, *Advanced Free Space optics (FSO): A Systems Approach*, Springer, New York, 2015.
2. A.C. Clarke, "Extra-terrestrial relays", Wireless World, October, 1945, pages 305-308.
3. Arun K. Majumdar, *Optical Wireless Communications for Broadband Global Internet Connectivity: Fundamentals and Potential Applications*, Elsevier, Amsterdam, Netherland 2019.
4. Y.-J. Li, S.-H. Zhao, J.-L. Wu, L. Liu et al, Designing of optical two-layered satellite network of LEO/MEO with global coverage, 2008 International Conference on Computer Science and Software Engineering, 12-14 Dec. 2008, Wuhan, China, **IEEE *Xplore*:** 22 December 2008
5. 'Metaverse': the next internet revolution? (2021, July 28) retrieved 29 July 2021 from https://techxplore.com/news/2021-07-metaverse-internet-revolution.html
6. Michael Ziemkiewicz, Scott R. Davis, Scott D. Rommel, Derek Gann, Benjamin Luey, Joseph D. Gamble, and Mike Anderson "Laser-based satellite communication systems stabilized by non-mechanical electro-optic scanners", Proc. SPIE 9828, Airborne Intelligence, Surveillance, Reconnaissance (ISR) Systems and Applications XIII, 982808 (17 May 2016); https://doi.org/10.1117/12.2223269
7. Hyosang Yoon, "Pointing System Performance Analysis for Optical Inter-satellite Communication on CubeSats," Doctor of Philosophy Thesis, MIT Dept. of Aeronautics and Astronautics, May 24, 2017.
8. David N. Amanor, William W. Edmonson, and Fatemeh Afghah," Inter-Satellite Communication System based on Visible Light," IEEE TRANSACTIONS ON AEROSPACE AND ELECTRONIC SYSTEMS, VOL. X, NO. X, XXXXX 2018, arXiv:1806.01791v1 [eess. SP] 2 Jun 2018.
9. Jose E. Velazco, Joseph Griffin, Danny Wernicke, John Huleis, Andrew DeNucci, Ozdal Boyraz, Imam Uz Zaman," Inter-satellite Omnidirectional Optical Communicator for Remote Sensing," Proc. SPIE 10769, CubeSats and NanoSats for Remote Sensing II, 107690L (18 September 2018); doi: https://doi.org/10.1117/12.2322367.
10. Christopher Hoy, Jay Stockley, Janelle Shane, Kelly Kluttz, Douglas McKnight and Steven Serati, "Non-Mechanical Beam Steering with Polarization Gratings: A Review," MDPI *Crystals 2021, 11,* 361. https://doi.org/10.3390/cryst11040361.
11. Tie Chi, Lixin Meng, Xiaoming Li and Lizhong Zhang, "Feasibility study of microsatellite laser communication pointing system based on SEEOR technology," Tie Chi et al 2019 J. Phys.: Conf. Ser. 1314 012015, doi:https://doi.org/10.1088/1742-6596/1314/1/012015.
12. Frantz, Jesse A, Myers Y, Jason D. Bekele, Robel Y, et al.," Chip-based nonmechanical beam steerer in the midwave infrared," Journal of the Optical Society of America B 2018,35(12) :C29-C37.
13. Peter M. Goorjian "Fine pointing of laser beams by using laser arrays for applications to CubeSats", Proc. SPIE 11678, Free-Space Laser Communications XXXIII, 116780E (5 March 2021); https://doi.org/10.1117/12.2575661.
14. T. Anfray *et al.*, "Assessment of the Performance of DPSK and OOK Modulations at 25 Gb/s for Satellite-Based Optical Communications," *2019 IEEE International Conference on Space Optical Systems and Applications (ICSOS)*, 2019, pp. 1-6, doi: https://doi.org/10.1109/ICSOS45490.2019.8978982.
15. Arslan Sajid Raja et al, Ultrafast optical circuit switching for data centers using integrated soliton microcombs, Nature Communications (2021). DOI: https://doi.org/10.1038/s41467-021-25841-8.
16. Sven Hochheim, Alexander Büttner, Peter Wessels, Jörg Neumann, Dietmar Kracht, Laser Zentrum Hannover e.V. (Germany)," 100W optical amplifier for 10 channel laser communication system with enhanced wall-plug efficiency in the 1μm wavelength range," Paper # 11993-18, SPIE Photonics West, 22-27 January 2022 San Francisco, California.

17. Paul Steinvurzel, Wei W. Chang, and Todd S. Rose, "Dual band optical amplifier for 1064 and 1550 nm communications," Paper # 11993-14, SPIE Photonics West, 22-27 January 2022 San Francisco, California.
18. Paul D. Shubert, Joshua Kern, and Pablo Reyes, "Free-space optical circulator for interoperable laser communications," Paper # 11993-10, SPIE Photonics West, 22-27 January 2022 San Francisco, California.
19. David Allioux, Antonin Billaud, Adeline Orieux, Fausto Gomez Agis, v, Kassem Saab, Stephane Bernard, Thibault Michel, Olivier Pinel, "Free space optical link demonstration using multi-plane light conversion turbulence mitigation," Paper 1993-4, SPIE Photonics West, 22-27 January 2022 San Francisco, California.
20. Karen Ravel, Charlie Koechlin, Eddie Prevost, Thierry Bomer, Romain Poirier, Laurence Tonck, Guillaume Guinde, Matthieu Beaumel, Nick Parsons, Michael Enrico, Sean Barker, "Optical Switch Matrix development for new concepts of Photonic based flexible Telecom Payloads," Proc. of SPIE Vol. 11180 111803H-1, International Conference on Space Optics — ICSO 2018, edited by Zoran Sodnik, Nikos Karafolas, Bruno Cugny.
21. Yasushi Munemasa, Yoshihiko Saito, Alberto Carrasco-Casado, Phuc V. Trinh, Hideki Takenaka, Toshihiro Kubo-oka, Koichi Shiratama, and Morio Toyoshima, "Feasibility study of a scalable laser communication terminal in NICT for next-generation space networks," ICSO 2018 International Conference on Space Optics, Chania, Greece 9-12 October 2018, Proc. of SPIE Vol. 11180 111805W-2.
22. Arun K. Majumdar, *Advanced Free Space Optics (FSO): A Systems Approach,*" Springer Science+Business Media, New York 1015.
23. Arun K. Majumdar, *Optical Wireless Communications for Broadband Global Internet Connectivity: Fundamentals and Potential Applications,*" Elsevier, Amsterdam, Netherlands 2019.
24. Timothy J. Brothers and Arun K. Majumdar, "Distributed sensing with free space optics," Chapter 14, in the book, *Principles and Applications of Free Space Optical Communications* by Arun K. Majumdar (Editor), Zabih Ghassemlooy (Editor), A. Arockia Bazil Raj (Editor), The Institution of Engineering and Technology (IET), UK, 2019.
25. PHUC V. TRINH, ALBERTO CARRASCO-CASADO, TAKUYA OKURA, HIROYUKI TSUJI, DIMITAR R. KOLEV, KOICHI SHIRATAMA, YASUSHI MUNEMASA, and MORIO TOYOSHIMA, "Experimental Channel Statistics of Drone-to-Ground Retro-Reflected FSO Links With Fine-Tracking Systems," in *IEEE Access*, vol. 9, pp. 137148-137164, 2021, doi: https://doi.org/10.1109/ACCESS.2021.3117266.
26. Ronald E. Meyers, Keith S. Deacon and Arnold D. Tunick, "*Free-Space and Atmospheric Quantum Communications,*" Chapter 10, in the Book, Arun K. Majumdar, *Advanced Free Space Optics (FSO): A Systems Approach*, Springer Science+Business Media, New York 1015.
27. https://en.m.wikipedia.org/wiki/Quantum_channel#
28. https://en.m.wikipedia.org/wiki/Quantum_network#
29. MIT Technology Review © 2021 (Martin Giles February 14, 2019).
30. Long-distance and secure quantum key distribution (QKD) over a free-space channel (2021, January 25) https://phys.org/news/2021-01-long-distancequantum-key-qkd-free-space.html Optics.org, News, 27 Oct 2021. Also see: Yuan Cao, Yu-Huai Li, Kui-Xing Yang, Yang-Fan Jiang, Shuang-Lin Li, Xiao-Long Hu, Maimaiti Abulizi, Cheng-Long Li, Weijun Zhang, Qi-Chao Sun, Wei-Yue Liu, Xiao Jiang, Sheng-Kai Liao, Ji-Gang Ren, Hao Li, Lixing You, Zhen Wang, Juan Yin, Chao-Yang Lu, Xiang-Bin Wang, Qiang Zhang, Cheng-Zhi Peng, and Jian-Wei Pan, "Long-Distance Free-Space Measurement-Device-Independent Quantum Key Distribution," Phys. Rev. Lett. 125, 260503 – Published 23 December 2020. Chen, YA., Zhang, Q., Chen, TY. *et al.* An integrated space-to-ground quantum communication network over 4,600 kilometres. *Nature* **589,** 214–219 (2021). https://doi.org/10.1038/s41586-020-03093-8
31. Carlo Liorni et al, " Satellite-based links for quantum key distribution: beam effects and weather dependence," 2019, New J. Phys. 21 093055.

32. R Alleaume´, F Treussart, G Messin, Y Dumeige, J- F Roch, A Beveratos, R Brouri–Tualle, J-P Poizat and P Grangier, "Experimental open air quantum key distribution with a single photon source," New Journal of Physics July 2004, DOI: https://doi.org/10.1088/1367-2630/6/1/092 · Source: OAI.

33. "A superconducting silicon-photonic chip for quantum communication," SPIE Publications 01 November 2021.

34. Xiaodong Zheng, Peiyu Zhang, Renyou Ge, Liangliang Lu, Guanglong He, Qi Chen, Fangchao Qu, LaBao Zhang, Xinlun Cai, Yanqing Lu, Shining N. Zhu, Peiheng Wu, Xiaosong Ma, "Heterogeneously integrated, superconducting silicon-photonic platform for measurement-device-independent quantum key distribution," Advanced Photonics, 3(5), 055002 (2021). https://doi.org/10.1117/1.AP.3.5.055002, 30 October 2021.

35. Sebastian Ecker, Bo Liu, Johannes Handsteiner, Matthias Fink, Dominik Rauch, Fabian Steinlechner, Thomas Scheidl, Anton Zeilinger and Rupert Ursin, "Strategies for achieving high key rates in satellite-based QKD," npj nature partner journals, Quantum Information (2021) 7:5; https://doi.org/10.1038/s41534-020-00335-5

36. Leong-Chuan Kwek1,2,3,4*, Lin Cao4,5, Wei Luo4,6, Yunxiang Wang7, Shihai Sun8, Xiangbin Wang9,10,11 and Ai Qun Liu," Chip-based quantum key distribution," Kwek et al. AAPPS Bulletin (2021) 31:15 https://doi.org/10.1007/s43673-021-00017-0

37. Milo Grillo; Alexis A. Dowhuszko; Mohammad-Ali Khalighi; Jyri Hämäläinen," Resource allocation in a Quantum Key Distribution Network with LEO and GEO trusted-repeaters, IEEE 6-9 Sept. 2021 17th International Symposium on Wireless Communication Systems (ISWCS).

38. Xingyu Wang, Wei Liu, Tianyi Wu, Chang Guo, Yijun Zhang, Shanghong Zhao and Chen Dong," Free Space Measurement Device Independent Quantum Key Distribution with Modulating Retro-Reflectors under Correlated Turbulent Channel," Entropy 2021, 23, 1299. https://doi.org/10.3390/e23101299

39. Cheng-Qiu Hu, Zeng-Quan Yan, Jun Gao, Zhan-Ming Li, Heng Zhou, Jian-Peng Dou, and Xian-Min Jin, Phys. Rev. Applied 15, 024060 – Published 24 February 2021

40. Zhao Feng, Shangbin Li, and Zhengyuan Xu," Experimental underwater quantum key distribution," Optics Express, Vol. 29, Issue 6, pp. 8725-8736, 2021.

41. L. Ji, J. Gao, A.-L. Yang, Z. Feng, X.-F. Lin, Z.-G. Li, and X.-M. Jin, "Towards quantum communications in free-space seawater," Opt. Express 25(17), 19795–19806 (2017).

42. Robert Bedington 1, Juan Miguel Arrazola1 and Alexander Ling, "Progress in satellite quantum key distribution," npj Quantum Information (2017) 0, www.nature.com/npjqi

43. Ivan Derkach1,*, Vladyslav C. Usenko1,†, and Radim Filip, "Squeezing-enhanced quantum key distribution over atmospheric channels," New Journal of Physics, Volume 22, May 2020.

44. Robert Bedington, Juan Miguel Arrazola and Alexander Ling," Progress in satellite quantum key distribution," npj Quantum Information, nature partner journals, 09 August 2017.

45. Teixeira, F.B.; Ferreira, B.M.; Moreira, N.; Abreu, N.; Villa, M.; Loureiro, J.P.; Cruz, N.A.; Alves, J.C.; Ricardo, M; Campos, R., "A Novel Simulation Platform for Underwater Data Muling Communications Using Autonomous Underwater Vehicles," Computers 2021, 10, 119., MDPI, https://doi.org/10.3390/computers10100119

46. Space: The Ultimate Network Edge, Data Storage Knowledge, October 17, 2016 http://www.datacenterknowledge.com/archives/2016/10/17/space-the-ultimate-network-edge

Chapter 8
Recent Demonstrations of Laser Satellite Communications: Path Forward for Constellation of Laser Satellite Technology for Global Communication

8.1 Introduction

Inter-satellite laser communication is one of the most efficient methods and has become a key technology for developing global satellite communication system to connect *anywhere*, *anytime* including remote locations. Using free-space optical communication (FSOC) with all-optical technologies that uses laser beams as carrier waves can transmit images, voices, and other information data in space very easily and efficiently. Laser communication using a single satellite at a given altitude is limited in covering all locations worldwide. The necessity of developing a constellation of laser satellites has been discussed in the previous chapters. Fundamentals and potential applications of broadband global Internet connectivity with optical wireless communications are described in the literature [1]. Laser space communications require two basic communication technologies: (1) inter-satellite laser communications within a given constellation at a certain orbital altitude and (2) inter-satellite laser communications between satellites belonging to different constellation groups. One of the important features of developing inter-satellite laser communication links is that it can reduce the dependence and design requirements of the satellite constellation system on the ground, terrestrial, and even underwater optical network optimizing the availability, cost, and reliability of the worldwide overall communication systems. The current space race of achieving space networking among low- and medium-orbit constellation projects is discussed in this chapter. Recently, a low Earth orbit (LEO) satellite communications company confirmed its successful deployment of 36 satellites, which now makes OneWeb's total in-orbit constellation to 394 satellites [2]. Another global satellite communication company, Mitsubishi Heavy Industries from Japan, launched Inmarsat-6s, the first ever dual-band hybrid L and Ka band satellite with increased capacity, new technological advances, and high-speed broadband capacity [3]. Note that these communication satellites operate in the RF ranges (L band, 1–2 GHz, Ka band, 27–40 GHz) and are *not* laser communication satellites. The Inmarsat

© Springer Nature Switzerland AG 2022
A. K. Majumdar, *Laser Communication with Constellation Satellites, UAVs, HAPs and Balloons*, https://doi.org/10.1007/978-3-031-03972-0_8

communication satellite will be raised to geostationary orbit (GEO) approximately ~22,500 miles above the Earth. The constellation of communication satellites will help toward the continuous demand from telecommunications providers and the OneWeb's goal of bringing improved digital communication services to some of the hardest-to-reach parts of the world. However, there are only very few actual demonstrations for laser satellite communications with lasers and optical/photonic component payloads in the space. One very recent demonstration example includes SpaceX Starlink "space lasers" test of Starlink satellite for the first time [4]. Out of ~715 Starlink satellites, SpaceX launched in the last couple of years only two spacecrafts as a part of Starlink-9 or -10 in August of 2021 that have successfully tested prototype *lasers in orbit* [4]. The absolute necessity of high-performance laser interlinks has been pointed out in a number of previous chapters of this book that justifies the design and developments of laser satellite constellations at various orbits and altitudes. Laser-based constellation of communication satellites can optimize user communications routed by laser to and from the ground stations physically located closest to the user and their traffic destinations. This chapter specifically addresses the demonstrations and the tests of laser-based satellites reported and discussed as of today and also includes recent demonstrations by the governments, industries, and research laboratories for establishing laser satellite communications for GEO, LEO, and small/CubeSats in various orbits and altitudes and thus to verify a series of on-orbit optical based technology. The goal is to develop the *path forward* for laser satellite constellations and establish convergence of aerial, terrestrial, underwater, and satellite networks.

Recent successful demonstrations and development of technology concepts for space-to-ground links, airborne links, and satellite-to-underwater optical communication links are discussed in this chapter along with the brief descriptions of the programs/projects of the big players in the space race. Some of the recent laser communication payload technologies essential for laser satellite communication systems are also described. The basics and the fundamentals discussed in the previous chapters of the book will be useful in understanding the various experiments, tests, and demonstrations, which paves the pathway to establish all-optical global Internet connectivity.

8.2   Relay-Assisted Free-Space Optical Communication to Extend Coverage Between Satellites: Improvement of Inter-satellite Performance

Relaying technique has been introduced in the literature as an effective method to extend coverage, which is essential in developing communication satellite constellations. Relay-assisted free-space optical (FSO) systems can also mitigate the atmospheric effects of fading during transmission and reception. A number of relaying techniques for free-space optical (FSO) communications are discussed addressing various communications scenarios (i.e., serial, parallel, or mesh) and applying different deployment protocols such as amplify-and-forward or decode-and-forward. Increasing inter-satellite communication links can put less dependence on the constellation, cost, and designs. This section mainly discusses the recent demonstrations and tests of laser satellite communications with a path forward to establish a constellation of laser satellites' technology eventually to provide solutions for achieving convergence of aerial, terrestrial, underwater, and satellite communication networks. In that respect, concept technology for developing all-optical relay-assisted FSO systems will be essential. To accomplish end-to-end communication and connectivity between terminals anywhere on the Earth, it needs a very complex design. This is because connecting space networks with various nodes as well as individual and other satellites moving in the constellation is extremely complex. In addition, some of the optical links need to be in a line-of-sight (LOS) configuration between satellites or other communication terminals in different orbits. Therefore, relays inserted in the link not only reduce the atmospheric induced turbulence or any other link loss but also extend link span and ensure higher link availability so that the overall system performance is improved. The problem of routing assignment is critically important in developing all-optical networks for laser satellite communications constellations. Efficient routing methods applied to optical space networks using various satellite constellations can provide potential solutions for meeting the demands of ever-increasing data requirements at high data rates. Satellites in various constellation orbits can solve the problem of establishing connectivity and Internet to the remote locations worldwide. The space optical networks which connect the source to the destination can include terrestrial, aerial, and space terminals. Relaying techniques for FSO communications can generally include various deployment configurations such as serial, parallel, or mesh connecting different satellite, space, and aerial terminals, each of which are equipped with optical transceivers capable of optically amplifying the signals and routing to the next terminals. If there is an obstacle such as a cloud between a satellite terminal and the optical ground station (OGS), the routing technique should be able to perform line-of-sight switching to another satellite by site diversity technology. An all-optical FSO relay-assisted system can mitigate the atmospheric turbulence-induced signal fading, extend the link distance, and improve link availability. All-optical relaying technique can also include different communication protocols such as amplify-and-forward (AF) and decode-and-forward (DF). Fundamental concept and basic concept technologies have been discussed recently by the researchers [5] to demonstrate the usefulness of the multihop relay-based FSO communication link for delivering medical services in remote areas. Similar techniques can be very well used for developing FSO-based satellite communications. A schematic of a multihop FSO link with serial relay configuration is shown in Fig. 8.1a, where the communication transmitter is represented as source (S) and receiver as destination (D), separated by two relay terminals ($R1$ and $R2$). A general architecture of bidirectional multihop relay-based FSO link illustrated in Fig. 8.1b also shows line-of-sight switching by site diversity techniques to overcome geographical obstructions. This basic architecture can be extended to apply to the laser satellite constellation,

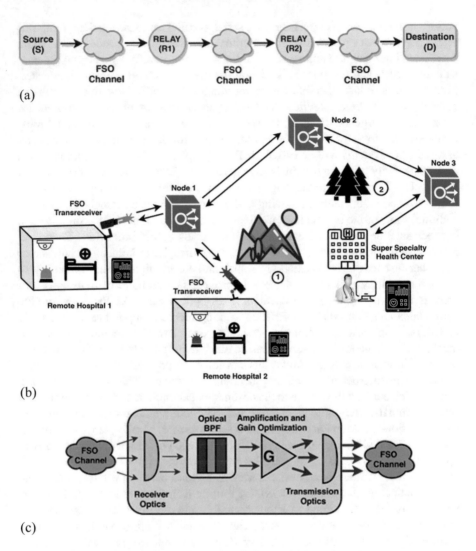

Fig. 8.1 (**a**) Relay-assisted all-optical multihop FSO link. (**b**) Concept technology for a potential healthcare system at remote centers through bidirectional multihop link. (**c**) All-optical relay terminal for the FSO links. See text for using this technology for exploring laser satellite constellation for developing space optical networks to connect optical ground stations to reach remote locations. (*Copyright Permission: Open Access—This work is licensed under a Creative Commons Attribution 4.0 License* [5])

establishing two-way FSO links with strategically placed optical ground stations in order to achieve high-data-rate communication and connectivity to remote locations worldwide. The research proposed in [5] investigated the operation of amplify-and-forward (A-F) terminal nodes based on all-optical multihop serial relay FSO link operating under fading channel. At each node, the signal processing is performed in the high-speed optical domain with no conversion to electrical or optoelectronic processing. To characterize the atmospheric turbulence, gamma-gamma (G-G) probability density function (PDF) of intensity fluctuations between nodes was assumed. The G-G form of the PDF is a generally accepted distribution function to be valid for weak to strong turbulence atmospheric conditions. As shown in Fig. 8.1c, the all-optical relay nodes used in the proposed link consist of specialized focusing lenses which collect the optical signal from the channel, a band-pass optical filter (BPF) to remove the background noise, and a gain-optimized optical amplifier (EDFA), where each node acts as an amplify-and-forward terminal. The specialized design tool *OptiSystem* was used in their research, which is particularly useful in maximizing power transfer with minimum noise amplification while using erbium-doped fiber amplifier (EDFA) action and amplified emission noise (ASE) leading to loss of coherence and depletion gain. Gain-flattening filter (GFF) optimization in the design tool *OptiSystem* was used in maximizing power transfer with minimum noise amplification. Transmission lenses at the other end of relay terminal form a coherent optical beam to propagate over free-space channel with minimum beam wander. The authors reported that multihop relay transmission can be designed to significantly improve the link performance and reliability, and the research demonstrated the potential for achieving 120 Gbit/s DP-16 QAM modulated multihop serial FSO link with coherent reception.

In their report, an example of the FSO signal at the destination node D for a twin-hop link is shown to be given by [5]

$$Y_D(t) = g_1 g_2 h_{R1R2} h_{R2D} \left(h_{SR1} S_0(t) + n_1(t) \right)$$
$$+ g_2 h_{R1R2} h_{R2D} \left(a_1(t) + n_2(t) \right) + h_{R2D} a_2(t) + n_3(t) \tag{8.1}$$

where

h_{SR1} = the channel gain for the link between S and $R1$.

$S_0(t)$ is the original signal transmitted by source node S.

$n_1(t)$, $n_2(t)$, and $n_3(t)$ are additive white Gaussian noises (AWGN) with zero mean and power spectral density N_0 at nodes $R1$, $R2$, and $R3$, respectively.

$a_1(t)$ and $a_2(t)$ = noises due to amplification of ADFA inside node $R1$ and inside $R2$ respectively.

h_{R1R2} and h_{R2D} = the channel gain for the link between nodes $R1$ and $R2$ and between nodes $R2$ and D, respectively.

g_1 and g_2 are the amplifier gains of nodes $R1$ and $R2$, respectively.

The received signal $Y_D(t)$ in the above equation can be generalized for any multihop FSO link with serial node configuration. Note that all-optical FSO link terminals can include laser satellite in a constellation such as LEO, CubeSat, small satellite, high-altitude platform (HAP), UAV, or aeroplane terminals equipped with laser/optical transceivers. The concept of multihop relay-based FSO communication link can thus deliver high-data-rate communication and connectivity in remote areas worldwide.

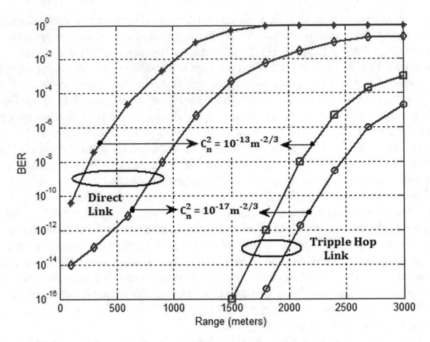

Fig. 8.2 All-optical communication performance of BER of direct and triple-hop FSO links for varying link ranges. (*Copyright Permission: Open Access—This work is licensed under a Creative Commons Attribution 4.0 License* [5])

Figure 8.2 shows [5] the BER performance for the proposed multihop FSO link operating in the 1550 nm window for different link ranges for different atmospheric turbulence conditions ($C_n^2 = 10^{-13}$ m$^{-2/3}$ (strong turbulence), and $C_n^2 = 10^{-17}$ m$^{-2/3}$ (weak turbulence)). Triple-hop FSO link results in superior error performance over double-hop and direct FSO links. The communication link ranges of 2440 m and 570 m were shown to be achievable for triple hop for BER = 10^{-5}. The proposed link could operate efficiently with optical power from -5 to 10 dBm. Finally, the design costs and complexities are critical in expanding convergence between optical networks and FSO links [5].

Their results demonstrated that all-optical amplify-and-forward multihop transmission can be considered as a promising solution for enhancing the performance of 120 Gbit/s DP-16 QAM FSO link for a variety of atmospheric channel conditions ranging from mild to severe. Similar techniques can be very well used for developing FSO-based satellite communications by integrating various laser satellites in a constellation and UAV relaying via reinforcement learning for non-terrestrial networks [6].

Recently, researchers reported in [7] all-optical amplify-and-forward (AOAF) protocol for investigating the performance of all-optical relaying system employing optical amplifier as the relaying technique. All-optical components in each relay of the basic AOAF relaying include transmitting and receiving optical lenses, optical fibers, and optical amplifiers. The two optical beams before and after the optical amplifiers act as transmitter and receiver for forwarding the beams to the next relay toward the destination. Their results show the performances of BER of a direct transmission and the dual-hop relaying system at the operating wavelength of 1550 nm using EDFA as optical amplifiers for two atmospheric turbulence strengths, $\sigma_R^2 = 0.2$ (weak) and 1.6 (strong): for example, direct AF system: (1) turbulence strength $\sigma_R^2 = 0.2$ (weak) for target BER ~10^{-5}, SNR (direct) ~27.5 dB, SNR (dual hop) ~21 dB, and (2) for $\sigma_R^2 = 1.6$ (strong), SNR (direct) ~32.8 dB, SNR (dual hop) ~26 dB. The results show the potential of the dual-hop all-optical relaying system compared to the direct link with no relay. The technique exploits the distance dependence of the turbulence effect while maintaining a high data rate [7].

The outage performance of relay-assisted FSO links with amplify-and-forward (AF) and decode-and-forward (DF) relaying schemes is discussed in [8]. The outage probability is relevant to laser satellite constellation for transmitting and receiving communication signals using various optical links. Their research paper addresses specifically the outage probability of communication signals for different relaying schemes. Amplify-and-forward (AF) relaying simply forwards the received signals to the next relay or the destination. In DF relaying, the relay at first decodes the signal after direct detection. Applying an appropriate modulation format, the modulated signal is then retransmitted to the next relay. By repeating this process, the signal finally reaches at the destination in a dual-hop scenario. In case of all-optical AF relaying, the signals can be processed in optical domain (using single-mode fiber ends for both collecting and transmitting the signals in conjunction with optical amplifiers (EDFA)) requiring the relay operation using low-speed electronic circuits to control and adjust the gain of amplifiers. This all-optical processing thus eliminates the EO/OE domain conversions to design and implement the optical transceiver system that is very efficient, fast, compact, and flexible. For a system using intensity modulation direct detection (IM/DD) employing binary pulse position modulation (BPPM) scheme, the outage probability of the serial AF relaying technique is given by [8]

$$P_{\text{out}} \approx Q \left(\frac{\ln\left(P_M^2 / (N+1)^2\right) - \mu_\varepsilon}{\sigma_\varepsilon} \right) \tag{8.2}$$

where N = number of relays, P_M = power margin = $\left(P_t^2 R^2 T_s^2 / N_0 \gamma_{\text{th}}\right)$, P_t = total transmitted power with average transmitted power of P, R = responsivity of the photodetector, T_s = duration of the signal/no-signal slots, μ_ε and σ_ε are the mean and variance of the normally distributed random variable ε, and the Q function is defined by $Q(x) = \left(\frac{1}{\sqrt{2\pi}}\right) \int_x^\infty \exp\left(-u^2 / 2\right) du$.

Figure 8.3 depicts the outage probability of serial relaying scheme for DF for an FSO communication system operating at a wavelength of $\lambda = 1550$ nm, where the end-to-end link range is $d_{S,D} = 5$ km, the atmospheric attenuation is 0.43 dB/km, and the strength of atmospheric turbulence $C_n^2 = 10^{-14}$ m$^{-2/3}$. The relay optimization location as discussed in [8] substantially improves the communication performance.

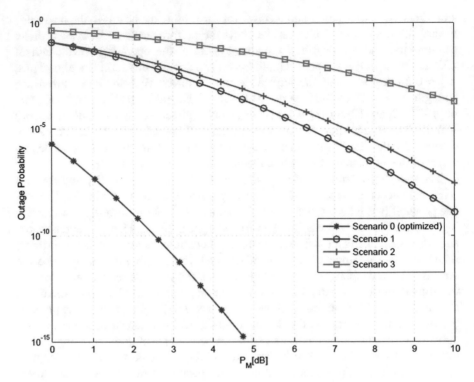

Fig. 8.3 Outage probability improvements using serial free-space optical (FSO) relay technique based on relay location optimization for a number of relays, $N = 3$. See Ref. [13]. Scenario 0 = (optimized): $d_1 = d_2 = d_3 = d_4 = d_{S,D}$ (source-to-destination distance)/4; scenario 1: $d_1 = d_4 = d_{S,D}$ (source-to-destination distance)/8, $d_2 = d_3 = 3d_{S,D}/8$; scenario 2: $d_1 = d_2 = d_3 = d_{S,D}/5$, $d_4 = 2d_{S,D}/5$; scenario 3: $d_1 = d_2 = d_3 = d_{S,D}/6$, $d_4 = d_{S,D}/2$ [8]. (*Reproduced with permission from* [8] Mohammadreza Aminikashani and Murat Uysal, "Relaying Techniques for Free Space Optical Communications" *published by The Institution of Engineering and Technology (IET), 2019*)

For example, for a target outage probability of 10^{-6}, the performance improvements in dB for three scenarios (see under the figure description) are 7.2, 8.2, and 13.4 dB compared to non-optimized scenarios. The result thus demonstrates the utility and usefulness of optimizing by serial AF relay technique. In practical laser satellite constellation and other space/ground optical links, the optimization technique reported in [8] can be applied between satellites, between satellites and space (e.g., HAPs, balloons, UAVs), and between terrestrial and optical ground-based terminals.

For a constellation of all-optical satellites, the outage probability is an important communication system performance parameter so that a reliable FSO communication system can be designed. There are various ways the relay can be placed between source and destination, and therefore the outage probability also depends on the atmospheric turbulence fading variance of the channel. The theoretical analysis in [8] illustrates the outage probability for full channel state information (CSI) and semi-blind all-optical AF relaying for different values of degree of freedom (DOF), i.e., M. For the ideal case, $M = 1$. Their results are shown in Fig. 8.4, which indicates

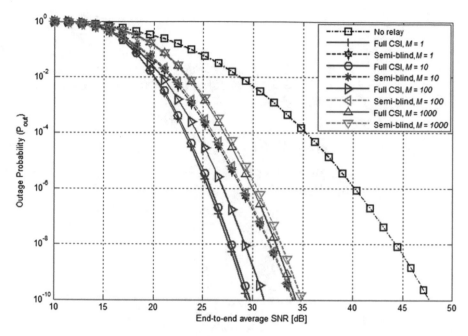

Fig. 8.4 Outage probability for all-optical AF relaying scheme using full-CSI and semi-blind relays and various degrees of freedom (DOF) of *M*-values. *M* = 1 indicates the ideal case, and *M* typically ranges from 100 to 1000 practical situations. (*Reproduced with permission from* [8] Mohammadreza Aminikashani and Murat Uysal, "Relaying Techniques for Free Space Optical Communications" *published by The Institution of Engineering and Technology (IET), 2019*)

a performance gain of 14.7 dB compared to the direct transmission for relay-assisted transmission at a target outage probability of 10^{-6}. The results are useful and applicable to the potential FSO laser communications between ground and GEO/LEO satellites, between satellites in a constellation, or between HAP/UAV and LEO scenarios in the presence of atmospheric turbulence-induced fading. Relay-assisted schemes have the potential of extending communication coverage mitigating the effects of fading in FSO links. Using the different relaying techniques, it will be possible to develop and design laser space networks to achieve the future goal of high-data-rate communication and connectivity worldwide.

8.3   Convergence of Aerial, Terrestrial, Underwater, and Satellite Communication Networks

Figure 8.5 summarizes different scenarios for developing all-optical worldwide connectivity providing high-data-rate communications connecting various optical ground stations (OGSs). Some of the typical communication links include inter-satellite, ground-to-satellite, remotely piloted aircraft (RPA) to LEO,

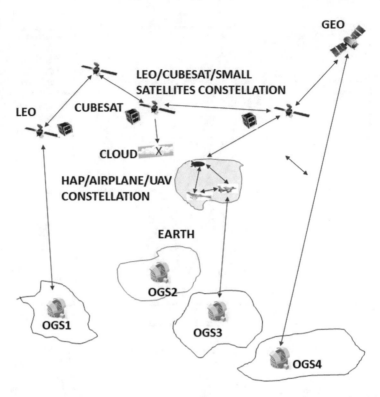

Fig. 8.5 Various scenarios and concepts of all-optical network worldwide connecting FSO links showing integrated system architecture for satellite-UAV/HAP/aerial-terrestrial terminals showing site diversity and redundancy

UAV-to-ground, satellite-to-ground, aircraft and GEO satellite, bidirectional inter-orbit, satellite-to-submarine, and GEO-GEO. Note that all the space and aerial links suffer from atmospheric effects at various levels based on the acceptable atmospheric profile models, for example, PAMELA [9], Hufnagel-Valley [10], and submarine laser communication (SLC) [11]. An excellent survey of atmospheric turbulence profiles is reported in [12], which includes some estimated C_n^2 profiles up to about 100 km. The estimated C_n^2 values can be extrapolated from [12] to be about much less than 10^{-19} m$^{-2/3}$. This means that for LEO or CubeSats for evaluating laser communications links, the atmospheric turbulence levels might be very small; however, for the GEO and LEO links to the grounds, the laser beam still has to suffer from the lower atmospheric turbulence effects (e.g., up to 20–40 km propagation path). Figure 8.5 depicts different scenarios for developing all-optical worldwide connectivity providing high-data-rate communications connecting various optical ground stations (OGSs). Some of the typical communication links include inter-satellite, ground-to-satellite, remotely piloted aircraft (RPA) to LEO, UAV-to-ground, satellite-to-ground, aircraft and GEO satellite, bidirectional interorbit, potential satellite-to-submarine, and GEO-GEO.

WHY do we need to increase the inter-satellite communication link distance?

- Put less dependence on the constellation, cost, and designs
- What are the factors involved: atmospheric propagation channel, laser power availability, receiver with low noise, and payload with very fast opto-electronic/optical technology components and processing capability
- Fine pointing, accuracy required and available (no moving part plus arrays of VCSELs, effects of moving terminals at each satellite)
- Relay-assisted technologies and the components/subsystems that can be installed
- Tx/Rx systems of each satellite belonging to inter-satellite links

8.4 Demonstrations and Technology Developments for Laser Satellite Communication Relevant to *Single* and *Constellation* of Laser Satellites

Free-space laser communications (FSLC) have the following advantages of using laser communications over radio waves: increased bandwidth (to transfer more data for enhanced communications) and less constraints in size, weight, and power (SWaP) requirements, all of which are attractive for designing and developing laser satellite communications (both single and a constellation). This section addresses mostly the recent demonstrations for laser-based satellite systems with reference to the related projects for FRLC in space for single and constellations of satellites in low Earth orbit (LEO) to provide global high-speed broadband connectivity and Internet access.

Small satellites are shown to be very useful in establishing the free-space optical communication (FSOC) links with optical ground stations (OGS). Some of the obvious advantages of small satellites are, for example, the following: CubeSats are typically $10 \times 10 \times 10$ cm cubes and can also be classified as picosatellites, nanosatellites, or microsatellites. The CubeSats typically weigh about 1 kg or slightly more of mass, use commercial off-the-the-shelf (COTS) components with very relaxed radiation requirements, and can also be accommodated as secondary payload in any launch vehicle or can be deployed from the International Space Station (ISS) as cargo. Therefore, the CubeSats are considered as one of the key enabling technologies for achieving free-space laser communications. CubeSats have the potential capabilities to establish lasercom as well as satellite communications. The NICT has been one of the pioneering institutes in successful demonstrations of lasercom technology, for example NICT's SOTA (Small Optical TrAnsponder), and is now involved in developing lasercom terminals with CubeSat platforms. Some of the recent research and development activities in these areas are described below.

8.4.1   NICT Demonstrations and Technology Developments

8.4.1.1   NICT Development of a Miniaturized Laser Communication Terminal for Very Small Satellites

The National Institute of Information and Communications Technology (NICT) has reported a current development of a miniaturized laser communication terminal, compatible with other different terminals, that can be easily installed in very small satellites to develop optical space network at low altitudes. The prototype will be installed in high-altitude platform systems (HAPs) to establish optical communication links between HAPs and optical ground stations (OGS), and eventually to geostationary (GEO) satellites to OGSs for potential world communication connectivity. One of the main goals of developing optically based terminals will be to satisfy the tremendous requirements of wireless communications for 5G and beyond applications. At present, RF-based communications exist which obviously cannot solve today's unprecedented data requirements of different applications of both data and videos. The NICT's approach to demonstrate various lasercom links using the developed lasercom terminals is discussed [13].

Examples of the NICT's potential applications and scenarios include HAPs-OGS, HAPs-OGS-GEO, HAPs-OGS-LEO, and HAPs-LEO-OGS. Eventually, small satellites such as CubeSats will be able to play important roles, and laser satellites belonging to constellations can all be optically connected to establish global broadband Internet connectivity, which can reach to any remote locations. CubeSat designed by the NICT was a 10 cm cube with a miniaturized Cassegrain telescope of 9 cm aperture. The space-qualified telescope exhibited a wave-front error of $\lambda/19$ RMS at the operating lasercom wavelength of 1550 nm and a total transmission of 93%. To amplify the laser signal optically, a specially designed miniaturized two-stage erbium-doped fiber amplifier (EDFA) was used, which has the capability of delivering 2 W of power during a lasercom link for LEO-ground scenario [13]. The NICT recently developed a number of miniaturized lasercom terminals to be integrated with a variety of space networks to support high-data-rate laser communications worldwide [14]. Some of these terminals are discussed by the researchers and include full transceiver and simple transmitter types [14]. The full transceiver type called FX terminal can support several thousands of kilometers in a LEO-LEO scenario in potential LEO constellations. The simple transmitter type called ST terminals is a further miniaturized version of FX terminal and is capable of supporting bidirectional communications at LEO-ground scenario. The NICT's lasercom prototypes of both the simple transmitter (ST) and full transceiver (FX) types are shown in Fig. 8.6. The designed lasercom terminal will finally be placed on board the ETS9 satellite to be launched in 2023 to the GEO orbit after analyzing and testing the full high-speed bidirectional power link budgets between GEO satellite and OGS. The demonstration of the prototype with HAP terminals is an important step toward developing worldwide optical wireless communications anywhere.

a. Simple transmitter type   **b. Full transceiver type**

Fig. 8.6 The NICT's first miniaturized laser communication terminal prototypes of (**a**) the simple transmitter (ST) and (**b**) the full transceiver (TX) types are shown. (*Courtesy of, and permission from Dr. Alberto Carrasco-Casado, NICT, 2022 to reuse* [14])

8.4.1.2   NICT Designs and Developments of Optical Ground Stations (OGS) for GEO and LEO Applications

A recent paper reviews and explains the recent advances made toward establishing wireless optical communication links for small satellites and constellations [15]. All these developments pave a pathway for future space laser communications to provide very-high-data-rate bidirectional worldwide communications satisfying small size, weight, and power (SWaP) constraints for installing payloads in the satellites. Effective inter-satellite inks are essential in establishing such space optical networks, which include spaceborne, near-space, aerial, terrestrial, and underwater terminals. Quantum cryptography and quantum key distribution (QKD) for secure laser communications have also been addressed in the report. A novel optical ground station to support satellite-to-ground links for both geostationary (GEO) and low Earth orbit (LEO) satellites has been discussed by the researchers at the NICT [16]. The latter is developing an innovative solution for atmospheric turbulence-induced compensation problem for ground-to-GEO uplink satellite laser communications as well as for downlink from GEO to ground when designing an efficient global laser-based satellite communication system.

Their proposed uplink wave-front pre-compensation solution and a picture of the developed optical bench are shown in Fig. 8.7. The NICT-proposed solution can be described as follows. In their approach, the wave-front information is obtained from the downlink light. For a point-ahead angle (PAA) of 20 μrad for a GEO communication scenario which includes about 20 km of atmospheric path affecting the laser beam, the distance between both the uplink and downlink paths can be about 40 cm. The researchers therefore considered the wave-front information for the whole telescope and not the data for downlink path only. The uplink wave front is then extracted using a propagation simulator and an algorithm. Laser beam aberrations for satellite-to-ground and uplink and downlink are therefore estimated with their proposed method. The compensated laser beams will then be used for two-way communications in the presence of atmospheric turbulence.

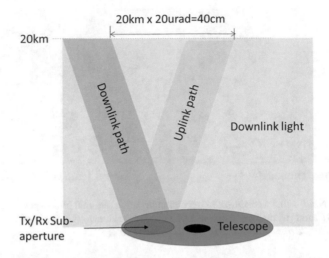

b.

Fig. 8.7 (**a**) Concept of the NICT's proposed uplink pre-compensation solution and (**b**) an actual picture of the developed optical bench. (*Courtesy of, and permission from Dr. Alberto Carrasco-Casado, NICT, 2022 to reuse* [16])

8.4.2   NASA's Satellite-Based Laser Communications Demonstrations

A recent payload launched aboard, the Space Test Program Satellite-6 that will demonstrate NASA's first two-way laser relay communications system, is discussed [17]. The Laser Communications Relay Demonstrations (LCRD), led by NASA's Goddard Space Flight Center in Greenbelt, Maryland, and partnerships with JPL, in Pasadena, California, and MIT Lincoln Laboratory in Massachusetts, will

Fig. 8.8 NASA's conceptual image of the laser communications relay demonstration payload transmitting optical signals. (*Credit: NASA Goddard Space Flight Center, LCRD, 2022* [17])

demonstrate the capabilities of sending and receiving data over invisible infrared lasers at about 10–100 times data rate than the existing RF systems traditionally used by the spacecraft. This demonstration will open the pathway toward developing optical based technologies for future satellite communications. LCRD will perform two-way free-space laser communications between a geosynchronous orbit (GEO) satellite and optical ground station (OGS) at 1.2 Gbit/s data. By 2030s, NASA expects to design and develop reliable and robust optical space communications infrastructure. Conducting future tests and experiments, NASA will also expect for future commercial missions using optical communications in Earth orbit and for exploration of the Moon, Mars, and beyond. Verifications of two-way laser communications by the LCRD mission will also serve as a relay between an optical communications terminal on the International Space Station (ISS) and ground stations on Earth [17] (Fig. 8.8).

8.4.3 Current Projects of Laser Communications for Satellite Constellations in LEO

A number of recent demonstrations and tests for exploring laser communication fully in space such as an inter-satellite laser link or a ground-to-satellite or satellite-to-ground application are summarized below:

Free-space laser communications (FSLC) technologies are very actively considered now by a number of corporations like SpaceX, Facebook, and Google for developing space networks, which include space and high-altitude platforms. One major application for exploring FSLC is to design and develop global broadband connectivity and satellite Internet. Various projects, concepts, communication terminals, scenarios, and achievable data rates are already explained and described in the literature [18] and therefore are not repeated in this section. Interested readers are encouraged to consult reference [18] for details. Some of the highlights of the projects emphasizing laser communication are summarized below:

- *European Data Relay System (EDRS)*/Tesat-Spacecom: GEO-LEO space-to-space, data rate 1.8 Gbit/s
- *Laser Light Communications*/Ball Aerospace & Technologies: MEO (medium Earth orbit) space-to space/ground, MEO, data rate 100 Gbit/s
- *SpaceLink*/Mynaric: Space-to-space MEO, LEO: Laser communication-enabled airborne and spaceborne communication networks, data rate 10 Gbit/s (see below for Mynaric* details)
- *EDRS-A*/SpaceDataHighway: A partnership between ESA and Airbus—the first satellite in global constellation—successful optical link to customer satellites for laser communication in space, uses secure laser communication from LEO

Mynaric: A laser communication company which develops and manufactures laser products for providing backbone connectivity to link aircraft, unmanned aerial vehicles (UAVs), high-altitude platforms (HAPs), satellites, and optical ground stations [19] and also to establish the necessary data highways for constellations and other mesh networks in air and space. Some of the applications and capabilities of Mynaric's products include laser communication in space covering space-to-space, space-to-air, and space-to-ground connectivity incorporating LEO and MEO satellites. However, note that the satellites need to be equipped with all-optical components such as laser transceivers with optical amplifiers for relaying as explained earlier. For example, the company's recent optical communications terminal suitable for next-generation constellations enables intra-plane and cross-plane inter-satellite connectivity in various orbits and optical links [19].

Small Optical Link for International Space Station (SOLISS): Establishment of bidirectional symmetric Ethernet links using laser communication devices designed for small satellites [20]. Three organizations in Japan, namely JAXA, NICT, and Sony CSL, demonstrated the successful bidirectional links between small optical link for International Space Station (ISS) and the optical ground station of NICT for satellite communications and transmitting high-definition (HD) image data via Ethernet [20]. This demonstration and successful tests clearly set a pathway for ultrahigh-speed and low-latency data communications for future cross-links between satellites and between satellite and ground station. The concept technology for laser satellite constellation located at various orbital heights and space/air and ground terminals can thus be designed and developed to establish two-way connectivity optically using laser links.

LUCAS System: NEC reported in [21] that it achieved inter-satellite data transmission rate of up to 1.8 Gbit/s, which is seven times higher than the existing system by developing laser communications terminals for satellites. The system called Laser Utilizing Communication System on board (LUCAS) basically enables data relaying between LEO sand GEO satellites. Two types of laser communications terminals were developed, one for GEO satellites and the other for LEO observation satellites. In order to develop the ultralong-haul laser communications over a distance of about 40,000 km between GEO and LEO satellites, NEC amplified low-output semiconductor laser light operating at the wavelength of 1.5 μm to high output under vacuum environment. Relay techniques discussed earlier have been utilized to send data via relay satellites in GEO as well as from LEO satellites to ground stations.

Some of the *future missions* for establishing laser communication in space including those currently under considerations are the following:

HICALI (High-speed Communication with Advanced Laser Instrument) lasercom terminal to be demonstrated in 2022–2023: The fastest bidirectional lasercom link between the geosynchronous orbit (GEO) and an optical ground station (OGS)—data rate of 10 Gbit/s—terminal on board the ETS-9 satellite and first inter-satellite links between a CubeSat in LEO and HICALI in GEO 1 year later [22]. LEO-to-GEO inter-satellite laser communication links will be beneficial to establish extremely high-data-rate communications between LEO and optical ground stations (OGS) by using relaying technique such as LEO-to-GEO-to-OGS. The average communication ranges for LEO-GEO and LEO-OGS scenarios are 39,693 km and 1103 km, respectively. A joint mission between NICT and the University of Tokyo will demonstrate this laser communication relaying technique between LEO CubeSat and GEO satellite. A 6U (340.50 × 226.30 × 100.00 mm) CubeSat will be used for 10 Gbit/s data transmission between LEO and GEO [23]. The HICALI terminal for performing bidirectional lasercom operates at a wavelength of 1550 nm using differential phase-shift keying (DPSK) modulation scheme, and the achievable data rate is 10 Gbit/s. A 1530 nm beacon for initial acquisition and to close the loop of the fine pointing system is required [23]. In order to close the LEO-GEO link with a diffraction-limited narrow laser beam, the point-ahead angle (PAA) also needs to be taken into account. This all-optical technological development of FSO via laser satellites clearly leads to a path forward for establishing high-data-rate communications using laser space networks to provide broadband connectivity to remote locations worldwide.

Integration of mega-satellite constellation and HAPs(UAVs) for non-terrestrial networks: In order to explore long-range communication between ground stations with low latency, recent research has been reported in [24] based on integrating huge number of LEO satellites in a mega-constellation with high-altitude platform (HAP) or UAV vehicle-assisted non-terrestrial networks. The researchers' proposed scheme uses deep reinforcement learning (DRL) with a novel action dimension reduction technique. The problem of forwarding packet between two faraway ground stations is discussed. The time-varying network topology using a large number of orbiting satellites and optimized HAP locations can thus create a three-dimensional (3D) wireless connectivity. The results in [24] address the practical proposed deployment communication scenarios (e.g., 1584 Starlink satellites at 550 km altitude, 2622 OneWeb satellites at 1200 km, and other similar proposed deployments). The simulation results are very useful and relevant to establish large-scale 3D wireless communication connectivity by seamlessly integrating LEO satellites, UAVs (HAPs), and ground terminals. Their work optimizes satellite association and HAP movement control to maximize the average data rate of long-range non-terrestrial under time-varying network topology. Their simulation parameters assumed the following: number of satellites in orbital plane = 22, distance between satellites = 1819 km, and bandwidth of RF link = 10^9 Hz. Note that the work in [24] considers RF communication. The method and the concept can be extended to apply to all-optical links with some modifications in the line-of-sight geometry for narrow laser beams and optical signal-to-noise ratio (SNR) taking into account optical path loss to reflect the multiple optical relaying techniques resulting in time-varying *optical* network topology. The future direction could be directed to explore the integration of mega-satellite constellation with UAVs or HAPs, all equipped with laser transceivers with *all-optical* components.

Flying Laptop satellite laser terminal to communicate with NICT's optical ground station: OSIRISc1 optical terminal on board the "Flying Laptop" was developed by DLR Institute of Communications and Navigation in cooperation with the Institute of Space Systems at the University of Stuttgart. Laser communication at the operating wavelength of 1550 nm was demonstrated at high data rate between OSIRISv1 terminal with no mechanical elements for beam steering and Japan's NICT optical ground station. The world's smallest terminal, OSIRIS-Cube1 in cooperation with Tesat-Spacecom, was launched on board the PIXL-1 satellite, and in the near future OSIRISv3 is expected to be installed on the Airbus Defence and Space Bartolomeo platform on board the International Space Station (ISS) [25].

Smallest laser terminal into orbit: In [26], DLR reported a compact satellite PIXL-1 that was successfully launched with the OSIRIS4CubeSat/CubeLCT laser terminal on board. The compact satellite carries the world's smallest laser terminal into orbit to test optical communications systems for various small satellites, which satisfy small size, weight, and power (SWaP) criteria to eventually connect HAPs (UAVs) and LEOs and optical ground stations (OGSs) to establish all-optical connectivity to remote locations for high-data-rate communications. The other big advantage of small satellites is their capability of functioning in mega-constellations, where a large number of satellites can be used for future laser-based small satellite communications for worldwide distribution of broadband Internet connection. Additional application for a constellation of small satellites for communications will be developing secure quantum communications by designing quantum key distribution between laser transmitters and receivers. Laser-based small satellites belonging to a constellation of small satellites can also be instrumental in building a worldwide network of optical ground stations to overcome the atmospheric challenges such as cloud-related obstacles for line-of-sight (LOS) laser links.

In a recent post [27], China's BeiDou GNSS reported to conduct an inter-satellite and satellite-ground station experiment using laser which could increase SatNav accuracy by a factor of 6–40 by synchronizing the satellites' atomic clocks with laser beams.

DARPA's Laser Communications Demo Satellites: In the Outreach news by DARPA [27], Defense Advanced Research Projects Agency (DARPA) announced the successful deployment of two satellites as part of SpaceX Transporter 2 launch for DARPA's Blackjack program to develop advanced laser communication technologies for the government. The on-orbit mission will be to demonstrate the viability of low size, weight, power, and cost of laser communication terminals that are interoperable for space architecture including inter-satellite links. Some of the companies involved in the design and development efforts include SEAKR Engineering, Advanced Solutions Inc., Lockheed Martin, and SpaceX. The Blackjack is an autonomous satellite constellation with a mix of commercial and military microsatellite constellations for space security.

The United States Air Force's funded program for developing laser terminal connecting military aircraft with satellites [28]: The program AFWERX is developing a pod to establish laser communication between aircraft and satellites in orbit. This Air-t-Space Laser Communications Pod could be deployed in a variety of USAF aircraft and placed under the wing of a fighter jet like the F-35 to transmit up to 10 Gbit/s data between airborne assets and communication satellites [28]. One of the hardest technological issues in achieving this goal is to develop the technology to maintain the pointing and navigating laser uplink during the maneuvering. Miniaturized adaptive optics (AO) can provide accurate pointing and tracking needed for the success of developing such optical communication terminals. If successful in future, other platforms such as high-altitude platforms (HAPs) can also be integrated with satellites in orbit and be a part of a mesh-like satellite network in a constellation of satellites. Recently developed miniaturized laser technologies using compact, fast optics and high-power lasers would make it possible in the near future to carry a high-power laser to reach satellites in orbit. Advanced tactical data links with an integrated architecture of laser space networks are therefore very much viable in the near future to establish high-data-rate communications between UAVs, HAPs, and other tactical aircraft with satellites in orbit. A constellation of 150 satellites is planned by 2024 to build a faster, more hardened military communications network based on laser links [28].

8.5 Summary and Conclusions

The technology concepts based on establishing the most efficient and reliable FSO laser links are explained in this chapter. The results and the current demonstrations of researchers from various laboratories and institutes clearly establish a pathway for developing potential worldwide communications capable of offering high data rates and high-capacity communications performance to *anywhere, anytime*. Small satellites such as CubeSats and microsatellites equipped with advanced optical/photonic components and devices will play an important role in establishing global connectivity and communications to satisfy the demands of constantly increasing high data rates. As a part of the solution, integration of a constellation of GEO, LEO, and CubeSats together with HAPs/UAVs with ground stations will definitely provide practical high-data-rate and high-capacity communications and connectivity all over the world. Advanced relaying techniques and recent successful demonstrations in various institutes and laboratories have established solid foundations to develop all-optical technology to reach remote locations. To achieve this goal, quantum secure communications will be required to develop long-distance quantum key distributions (QKD) for the FSO communication links. Finally, if successful, this technology will eventually solve the problem of *A Light in Digital Darkness: Optical Wireless Communications to Connect the Unconnected.*"

References

1. Arun K. Majumdar, *Optical Wireless Communications for Broadband Global Internet Connectivity: Fundamentals and Potential Applications*, Elsevier, Amsterdam, Netherland 2019.
2. https://oneweb.net/media-center/oneweb-confirms-successful-launch-of-36-satellites-after-rapid-year-of-progress
3. https://gcaptain.com/inmarsat-launches-worlds-most-sophisticated-communications-satellite/
4. https://www.teslarati.com/spacex-starlink-space-lasers-first-orbital-test/
5. Rajan Miglani, Jagjit Singh Malhotra, Arun K. Majumdar, Faisel Tubbal, Raad Raad, "Multi-Hop Relay Based Free Space Optical Communication Link for Delivering Medical Services in Remote Areas", DOI https://doi.org/10.1109/JPHOT.2020.3013525, IEEE Photonics Journal, Vol. 12, No. 4, August 2020.
6. Ju-Hyung Lee, Jihong Park, Mehdi Bennis, and Young-Chai Ko, "Integrating LEO Satellite and UAV Relaying via Reinforcement Learning for Non-Terrestrial Networks," arXiv:2005.12521, https://arxiv.org/abs/2005.12521.
7. Norhanis Aida Mohd Nor, Matej Komanec, Jan Bohata, Stanislav Zvanovec, Zabih Ghassemlooy, Manav R. Bhatnagar, Mohammad-Ali Khalighi, "All-Optical relay-assisted FSO systems" in the book. IET London, UK 2019, *Principles and Applications of Free Space Optical Communications,* Edited by Arun K. Majumdar, Zabih Ghassemlooy and A. Arockia Bazil Raj.
8. Mohammadreza Aminikashani and Murat Uysal, "Relaying Techniques for FreeSpace Optical Communications," in the book. IET London, UK 2019, *Principles and Applications of Free Space Optical Communications,* Edited by Arun K. Majumdar, Zabih Ghassemlooy and A. Arockia Bazil Raj.
9. E. Oh, J. Ricklin, F. Eaton, C. Gilbreath, S. Doss-Hammel, C. Moore, J. Murphy, Y. Han Oh, and M. Stell, "Estimating atmospheric turbulence using the PAMELA model," Proc SPIE, Free Space Laser Comm. IV, vol. 5550, pp. 256–266, 2004.
10. A. Majumdar and J. Ricklin, Free-Space Laser Communications: Principles and Advances, vol. 2. Springer, 2008.
11. H. Hemmati, ed., Near-Earth Laser Communications. CRC Press, 2009.
12. Lucien Canuet, "Atmospheric turbulence profile modeling for satellite-ground laser communication," Master's Thesis, Universitat Politècnica de Catalunya Master in Aerospace Science & Technology July 2014.
13. Alberto Carrasco-Casadoa, Koichi Shiratamaa, Phuc V. Trinha, Dimitar Koleva, Yasushi Munemasaa, Yoshihiko Saitoa, Hiroyuki Tsujia, Morio Toyoshima, "Development of a miniaturized laser-communication terminal for small satellites," 72nd International Astronautical Congress (IAC), Dubai, United Arab Emirates, 25-29 October 2021, IAC-21-B2.2.

14. Alberto Carrasco-Casado, "Free-space Laser Communications for Small Moving Platforms", Optical Fiber Communication Conference (invited paper), San Diego (United States), 6-10 March 2022.
15. Morio Toyoshima, "Recent Trends in Space Laser Communications for Small Satellites and Constellation", DOI https://doi.org/10.1109/JLT.2020.3009505, Journal of Lightwave Technology
16. D. R. Kolev, K. Shiratama, H. Takenaka, A. Carrasco-Casado, Y. Saito and M. Toyoshima, "Research and development of an optical ground station supporting both GEO- and LEO-to-ground links," *Advances in Communications Satellite Systems. Proceedings of the 37th International Communications Satellite Systems Conference (ICSSC-2019)*, 2019, pp. 1-9, doi: https://doi.org/10.1049/cp.2019.1248.
17. https://www.nasa.gov/mission_pages/tdm/lcrd/index.html
18. https://en.m.wikipedia.org/wiki/Laser_communication_in_space#
19. https://mynaric.com/
20. "Small Optical Link for International Space Station (SOLISS) Succeeds in Bidirectional Laser Communication Between Space and Ground Station" (https://global.jaxa.jp/press/2020/04/20200423-1_e.html). JAXA. 23 April 2020. Retrieved 7 August 2020.
21. "LUCAS" (https://www.satnavi.jaxa.jp/project/lucas/). JAXA. 30 October 2020. Retrieved 29 November 2020.
22. https://en.m.wikipedia.org/wiki/Laser_communication_in_space#
23. A. Carrasco-Casado *et al.*, "Intersatellite-Link Demonstration Mission between CubeSOTA (LEO CubeSat) and ETS9-HICALI (GEO Satellite)," *2019 IEEE International Conference on Space Optical Systems and Applications (ICSOS)*, 2019, pp. 1-5, doi: https://doi.org/10.1109/ICSOS45490.2019.8978975.
24. Ju-Hyung Lee, Jihong Park, Mehdi Bennis, and Young-Chai Ko, "Integrating LEO Satellite and UAV Relaying via Reinforcement Learning for Non-Terrestrial Networks," arXiv:2005.12521, https://arxiv.org/abs/2005.12521.
25. DLR German Aerospace Center, News 25 March 2021, DLR laser terminal in space establishes contact with Japanese ground station, https://www.dlr.de/content/en/articles/news/2021/01/20210325_dlr-laser-terminal-in-space-establishes-contact-japanese-ground-station.html
26. https://www.reddit.com/r/cubesat/comments/l7zts9/the_pixl1_satellite_was_just_successfully/ (Posted by u/Aerothermal) https://www.dlr.de/content/en/articles/news/2021/01/20210124_pioneering-launch-compact-satellite-with-smallest-laser-terminal.html
27. Inside GNSS, December 2, 2021 https://insidegnss.com/beidou-conducts-laser-communication-experiment-steps-ahead-of-u-s-could-improve-satnav-accuracy/
28. DARPA Outreach 7/7/2021 https://www.darpa.mil/news-events/2021-07-07

Chapter 9
Summary, Conclusions, and Future Directions

9.1 Introduction

The major contribution of this book is to introduce and analyze the fundamental concepts of *all-optical* technologies to design and construct optical satellite network as part of a larger integrated space/terrestrial network. Recent developments of optical/photonic devices and components have the potential realization of an optical satellite network of global extent space system architectures using the network. Laser satellite developments provide a future potential solution to establish high-data-rate global connectivity and communication using space optical networks, which include LEO and GEO satellites, aerial high-altitude platforms (HAPs), UAVs, CubeSats, and terrestrial and underwater optical terminals. Practical working communication systems integrating all these terminals do not still exist today since most of the radio frequency (RF)-based devices and systems have been the prominent and mature technology to address these. RF-based technologies and existing systems cannot solve future high-data-rate worldwide connectivity and communication problems. All-optical technology concept as addressed in this book is the only potential solution to reach these goals. Readers should also note that there are no off-the-shelf (OTS) optical wireless communication systems for addressing different scenarios available today. The materials in this book provide a systematic pathway to address the future global communications scenarios and discuss the technologies required to establish the goal.

The motivation of creating such a book started when the author was invited to be a panelist at the Optical Society of America (OSA) Optics and Photonics Congress conference on *"A Light in Digital Darkness: Optical Wireless Communications to Connect the Unconnected,"* in California, July 2019. The *digital divide* exists for the people without even any reliable Internet connection. High-altitude platforms (HAPs)/UAVs and satellite constellation (GEO, LEO, CubeSats) backhauls to connect with the ground stations seem to be the most viable solution for establishing worldwide connectivity in remote locations. In this respect, laser satellite with site

© Springer Nature Switzerland AG 2022
A. K. Majumdar, *Laser Communication with Constellation Satellites, UAVs, HAPs and Balloons*, https://doi.org/10.1007/978-3-031-03972-0_9

diversity and future optical wireless mesh network seems to be the best practical solution.

This book is an attempt to respond to these technological needs and developments for satisfying the ever-increasing demand for the data traffic worldwide driven by the ever-increasing number of wireless broadband Internet, mobile phones, smart devices, social web, and other video game applications [1]. The International Telecommunication Union (ITU) just published new data of facts and figures in 2021 revealing strong global growth in Internet use with the Internet users surging to 4.9 billion in 2021, while an estimated 37% of the world's population or about 2.9 billion people still never used the Internet [2]. This has put a tremendous pressure on the network infrastructure. Also, the number of mobile and wireless users together with data traffic is predicted to increase thousandfold over the next decade, which results in the mobile spectrum congestion (i.e., bandwidth bottleneck) at both the backhaul and last-mile access networks [3]. This book discusses the development of all-optical technology concept for satisfying the demand for high-bandwidth and secure communication using free-space optical (FSO) communications. Renewed interest in laser communications (i.e., lasercom) and potential of laser satellite communications for establishing broadband connectivity and satellite Internet are clearly providing the pathway for creating space networks, which can include space/aerial/terrestrial and underwater terminals. Some of the highlights for developing all-optical technology concepts necessary to achieve broadband global connectivity as well as the issues/consequences and the need for future direction are briefly summarized.

9.2 Summary of this Book's Major Contributions

- The main emphasis of this book is the realization and development of optical satellite network of global extent using *all-optical technology*. Space system architectures using the optical satellite network can establish high-data-rate communication and connectivity from anywhere to anywhere worldwide and can be economically viable. The laser links include ground, terrestrial, aerial, and space/satellite terminals equipped with laser/optical transceivers [4, 5]. Underwater optical terminals have also been included even with satellite-to-submarine laser communication links.
- The book has illustrated various scenarios for the deployments in LEO satellite-to-ground links (SGL) under various atmospheric turbulence conditions. Some of these scenarios include constellations of LEO/CubeSats interconnected by optical cross-links and connected to terrestrial user terminals and interconnected with fiber via "gateways" forming a global heterogeneous network.
- The book has assessed the ability of free-space optical communication (lasercom) to identify key optical technology developments and address the predicted performance.

- It has also emphasized laser links to be the most promising solution for bottle-necks in satellite-to-ground communication links (SGL).
- Time is just about right to integrate the advanced technologies and the laser satel-lites including CubeSats to satisfy severe size, weight, and power (SWaP) constraints.
- The book aims to help develop design goals for innovative technology concepts for direct high-data-rate (10s and 100s of Gbit/s) LEO-to-ground downlink, LEO/HAPs/UAVs to GEO, and GEO to ground for future applications.
- It has also explored the most recently developed advanced optical/photonic com-ponents/subsystems such as extremely compact, power-efficient optical trans-mitters (i.e., VCSEL array) and silicon-photonic devices for precise pointing with no moving parts.
- Novel technique of mitigating atmospheric turbulence using machine-to-machine (M2M) applicable to both terrestrial and space FSO communication was introduced.
- The book has explored the components of an inter-satellite laser link system for large satellites to establish cross-links as a means of establishing communication and the feasibility of a system to achieve long-distance cross-links for small sat-ellites (CubeSats).
- It has also discussed novel optical materials and devices (fast, miniaturized, SWaP, etc.), which can be integrated with the satellite and aerial terminals for different communication scenarios.
- It has developed quantum communications for satellite and terrestrial based ter-minals operating under atmospheric turbulence conditions—explaining the gen-eration of quantum key distribution (QKD).
- Relay-assisted FSO systems have been introduced for the first time as an effec-tive method to extend coverage semiofficially for ground-LEO-GEO-ground laser links, which can even include a constellation of laser satellites in different orbits. In particular, amplify-and-forward serial relaying with all-optical schemes has been discussed for establishing global connectivity to remote locations *any-where, anytime.*
- It has addressed future 5G/6G communication technologies.
- The book has also discussed applications for global Internet connectivity.

9.3   Impact and Issues of Existing and Planned Large Satellite Constellation for Communications

While constellation of communications satellites can provide potential technical solution for establishing global communication and connectivity as well as are designed to provide communication services to undeserved and remote areas, satel-lite constellations impact various space risks. The sustained growth in future satellite constellations due to the continuous increased number of active satellites in orbit has potential consequences in space as discussed below:

1. Satellite operations at risk from collision with each other or with orbital debris
2. Space debris impact: the hazards from orbital debris
3. Interference with astronomical observations affecting science programs that require twilight observation: add to the light pollution
4. Radiation exposure and atmospheric satellite drag

9.3.1 Space Debris and Space Junks

Space debris is a growing problem specially when they are floating in any orbit. Space junks are objects ranging from paint flakes from the space station to larger objects like spacecrafts floating in the orbits [6, 7]. The European Space Agency (ESA) recently reported that Space Surveillance Networks track about 29,210 pieces of debris on a regular basis [7]. There is thus a constant risk of serious damage to a functioning spacecraft due to potential slams of objects with other objects in the orbits at speeds of up to 5 miles/h. A commercial space monitoring company, ExoAnalytic Solutions, also reported in [6] that a defunct navigation satellite (a dead satellite) was pulled by a Chinese satellite on January 22, 2022, so that the space debris from the Earth's geosynchronous orbit (GEO) was taken away. Another space debris in 2021 hit the Canadian-made robotic arm used on the International Space Station (ISS) and suffered a "limited" damage [6]. Recently, EurAsian Times also reported [8] about the maneuverable vehicle being developed by Russia that can deorbit space debris, and it might be possible to deorbit large space junk objects. In future, the maneuverable vehicles will be engineered to drop space junk from orbit to be burned in the atmosphere or to fall safely on Earth. GEO satellites should be moved out, preferably their orbits, to be raised by about 300 km of their geostationary ring after their space mission to avoid any collision risk and interreference with active GEO spacecrafts. Space agencies need more resources into space debris mitigation for future operations and successful science missions and to be the most cost-effective solution for operating space satellites. Space debris can also be caused by explosions in space or missile tests conducted to destroy satellites, and typically they float in low Earth orbit (LEO) at speeds around 15,700 miles/h. Satellites or spacecrafts therefore have a good chance to collide with the debris materials and can be damaged. When very inexpensive communication satellites are deployed in very low orbits (for example below 600 km) to reduce latency, the satellites also suffer this consequence of causing collisions and space debris and thus causing hazards from orbital debris.

9.3.2   Space Laws Relevant to Deployed Communication Satellites

The United Nations Committee on the Peaceful Uses of Outer Space (UNCOPUOS) oversees the five international treaties. The Outer Space Treaty supports the principles governing the activities of the states in the exploration and use of outer space, including the Moon and other celestial bodies [9]. The treaty presents principles for space exploration and operation, which include deployment of space communication satellites. The international space law supports space activities for the benefits of all nations who are also responsible for damage caused by their space objects to avoid contaminating space and celestial bodies. As presented in the principles, *"No nation can 'own' space."*

9.3.3   Space Junks Affecting the Environment

Thousands of particles of space trash are generated by collisions of the objects in the orbits. Space junk can have detrimental effects on Earth's environment. A large number of debris generated by collisions between active and inactive satellites enter the Earth's atmosphere and can remain in orbits for at least a century or some debris can remain in orbits for at least 20–30 years. Obviously, these are the sources for *space pollution.* Slowly, a portion of the space junk in low Earth orbit will lose altitude resulting in being burnt up in Earth's atmosphere and will again have detrimental effects on the environment we are living and in causing health-related complications. Proposed mega-constellations and developments of miniature satellites such as CubeSats are expected to increase these global environmental problems.

9.3.4   Astronomical Observations Affecting Science Programs

Many astronomical observations require twilight observations for accurate spectral imaging of distant objects such as searching Earth-threatening asteroids and comets, outer solar system objects, and gravitational wave source. The Sun can be below the horizon for observing objects from the ground, but the satellites in a constellation hundreds of kilometers overhead are illuminated, which affects and limits the astronomical observations. Only if satellites are below 600 km a limited astronomical observation can be made during the night's darkest hours. However, the planned constellation of satellites orbiting at 1200 km may be visible mostly all night. The optical observatories will therefore have serious negative consequences to conduct astronomical observations for fundamental research. The noise photons due to the brightness from the constellation satellites overhead will simply be too high to produce any high-resolution imaging even with the most advanced optics and detectors existing today in any advanced optical observatory.

9.3.5   *Internet-Beaming Mega-Constellations Affecting Search for Life on Distant Planets*

SpaceX's Starlink and other mega-constellation satellites can interfere with the received signals of radio telescopes involved in investigating the search for life [10]. Readers should note that this subsection deals with the radio spectrum of radio astronomy observatories relevant to broadband Internet constellation. Besides the research on searching for life in other planets, the mega-constellations can also affect the research for hunting for *exoplanets* and the study of the most distant galaxies and planets in the universe. Cellular telephony and wireless Internet use the same wavelengths of the weak radio signals emitted by the stars and planets. The desired band for search for life and distant planets lies between 10.7 and 12.7 GHz, which is used by the Starlink satellites to beam Internet. This causes an interference for observations by the world's even largest radio telescope. Readers should <u>note</u> that the proposed concepts of *all-optical* technologies discussed in this book are not directly related to this because optical wavelengths are not affected by electromagnetic interference (EMI) as RF wavelengths. For the search of life or distant planets and galaxies, different optical bands need to be chosen to probe and sense the optical signals emitted by the distant planets.

9.3.6   *Concerns for Earth-Orbiting Satellites: Radiation Exposure and Atmospheric Satellite Drag*

The two main space weather-related concerns are the radiation exposure and atmospheric satellite drag [11]. Satellites are all the time exposed to the atmosphere. Because of the interaction of charged particles and electromagnetic radiation with the satellite's surfaces, and photonic and electronic instruments and components, the satellites can be damaged to function properly. In addition, atmospheric drag can also have significant impact on LEO satellites where the air resistance in those layers of the atmosphere is strong enough to pull the satellites closer to the Earth. The atmospheric drag thus basically perturbs the satellites' original intended orbits.

9.4   Conclusions and Future Directions

There is yet no practical communication system based on all-optical technologies to provide worldwide high-data-rate connectivity to any remote location integrating space, terrestrial, and underwater terminals. The satellites are not equipped with optical/photonic transceivers and modules to develop space optical networks to achieve this goal. However, research and development in worldwide laboratories, universities, and institutes are constantly being performed to design and develop advanced optical/photonic devices and modules necessary to establish broadband global Internet and communications connectivity.

Based on the recent developments of advanced high-speed FSO communication links, it seems to be feasible to construct optical satellite network (OSN), which is practically a part of a larger integrated space/terrestrial/underwater network. Demands for growing increase in the data traffic due to the widespread use of smart devices anytime, anywhere can be solved using all-optical technology-based construction of integrated space/terrestrial/underwater network. Optical space communications definitely will provide an important building block for future wide-area space data networks. This is what essentially this book aims to provide as one of the most coherent and comprehensive sources available today to describe the technologies' concepts and architectures required to design space/aerial/terrestrial/underwater connectivity and communication to reach to remote locations worldwide. High-fidelity analog transmissions and satellite-borne processing centers would be useful to develop new space system architectures and extremely high-data-rate network. The particular emphasis placed in this book is mainly on the use of *satellite constellation* and *inter-satellite* links for completing the space system architectures and data networks with all-optical technology concepts. RF-based technologies with spectrum congestion cannot be used or are not suitable to solve this mission, and optical technology is the only possible solution. All-optical relaying techniques to extend coverage and mitigate the effects of fading due to atmospheric turbulence for various deployment configurations, as described in this book, also support these. When the architectures as outlined and described in this book are fully exploited, the optical satellite network will become *economically viable* as well as will be able to offer extraordinary services to bring home the breakthrough in optical satellite networking. In short, there are many future opportunities for designing and creating the new dimensions of space system architectures to support this integrated network. It is hoped that this book will serve as a starting point for new developments of worldwide connectivity and communications using all-optical space networks with constellation of laser satellites.

Some of the challenges and future research where optical/laser technologies will play a significant role in future include the following:

- Need for optimized topology for integrated space, terrestrial, and underwater terminals for establishing all-optical global connectivity and communications and powerful fast processing tools: A combination of FSO terminals and high-speed fiber-optic backhaul would be a practical solution.
- Need for developing compact, attachable optical/photonic transceiver devices for developing comprehensive space optical networks: fast, silicon-based photonic devices and components are required.
- While low-power laser transmitters are required for small/CubeSats to send/receive data between satellites or between small satellites and ground stations, higher power laser transmitters are needed for HAPs/UAVs-to-LEOs, LEOs-to-GEOs, and GEOs-to-ground stations.

- There is a need to investigate the issues and consequences of mega-constellation satellites.
- Quantum communication is still the highest level of security. Efficient techniques for generating various quantum key distribution (QKD) schemes for long-distance atmospheric propagation will be useful.
- FSO technology will be a critical enabler for space-based Internet communication in which problems associated with controlling optical networks need to be solved.
- More efficient than existing low-power-consumption units with zero maintenance with minimum performance degradation and very long mean time between failures will be needed. Breakthrough digital payload electronics for LEO/CubeSats capable of low-noise, high-density-power delivery networks will be required.
- Real-time, dynamic optical technology and time-varying topology for steering of laser transceivers will be required. This will also address to avoid any routing oscillation between space and ground networks.
- There is a need to design and develop practical space-based quantum communication and quantum key distribution (QKD) for the new-generation optical receiver/transmitter devices required for long-distance atmospheric links between space, terrestrial, and underwater terminals. This should help improve security for laser satellite constellation-based communication.
- In future, satellite laser terminals should be extremely flexible to be able to communicate seamlessly with any space network, dynamically optimizing laser links between them.

References

1. Zabih Ghassemlooy, Arun Majumdar, and Arockia Bazil Raj, "*Introduction to free space optical (FSO) communications,*" Chapter 1 in the Book, "*Principles and Applications of Free Space Optical Communications,*", published by the Institute of Engineering Technology (IET), London, UK, 2019, Edited by Arun K. Majumdar, Zabih Ghassemlooy and A. Arockia Bazil (Raj).
2. ITU News. 29 Nov 2021-Facts and Figures 2021: 2.9 billion people still offline [Online] https://www.itu.int/hub/2021/11/facts-and-figures-2021-2-9-billion-people-still-offline/
3. Arun K. Majumdar, "Advanced Free Space optics (FSO): A Systems Approach," Springer, New York 2015.
4. Arockia Bazil Raj, Arun K. Majumdar, "Historical perspective of free space optical communications: from the early dates to today's developments," IET Communications, IET Commun., 2019, Vol. 13 Iss. 16, pp. 2405-2419, doi: https://doi.org/10.1049/iet-com.2019.0051.
5. Arun K. Majumdar, *Optical Wireless Communications for Broadband Global Internet Connectivity: Fundamentals and Potential Applications*, Elsevier, Amsterdam, Netherland 2019.
6. ExoAnalytic Solutions: https://exoanalytic.com/space-domain-awareness/ Universe Today-Space and astronomy news, February 1, 2022 by Evan Gough.
7. Earth.Org Sep 6th 2021: https://earth.org/space-junk-what-is-it-what-can-we-do-about-it/?gclid=EAIaIQobChMIsOHW36OC9gIVjSlMCh098wPkEAAYAyAAEgJNn_D_BwE

8. The Eur Asian Times, February 3, 2022.
9. Space Foundation Editorial Team, Report, 2022: Space Briefing Book-Space Law: International Space Law.
10. Space.com Report-published February 2022: "Mega-constellations like SpaceX's Starlink may interfere with search for life by world's largest radio telescope".
11. https://www.nbcnews.com/science/space/spacex-says-40-starlink-satellites-lost-geomagnetic-storm-rcna15516

Index

© Springer Nature Switzerland AG 2022
A. K. Majumdar, *Laser Communication with Constellation Satellites, UAVs, HAPs and Balloons*, https://doi.org/10.1007/978-3-031-03972-0

Printed in the United States
by Baker & Taylor Publisher Services